Praise for *Your Brain on Ink*

"In this companionable and lively workbook, the complex worlds of positive psychology, neuropsychology and expressive writing weave together. The result is accessible information and practice—an amazing feat of integration and utility.

Deborah Ross and Kathleen Adams make this weave look simple like the Russian Ballet makes turns and leaps look easy. The workbook clearly explains the neurological context of self-improvement and the expressive writing practice that is its vehicle. The book moves gracefully between information, exercises, and dialogs; the reader/writer is engaged, active, and absorbed throughout. *Your Brain on Ink* presents state-of-the-art information and writing techniques in a fun and practical way. It is a valuable contribution to the journal-writing world."—**Beth Jacobs**, PhD, author of *Writing for Emotional Balance* and *Paper Sky: What Happened After Anne Frank's Diary Ended*

"Ross and Adams have hit a home run with this workbook—it's compelling, informative, and very easy to read. The authors combine advice and insights about the power of therapeutic writing with the scientific research to prove why this kind of writing is so important. Interactive exercises and activities allowed me to integrate the content on a much deeper level. Now I'm even more committed to my journaling! I will return over and over again." —**Robert MacPhee**, former director of training for Jack Canfield; creator, Excellent Decisions Leadership Program; author of *Manifesting for Non-Gurus* and partner journal

"*Your Brain on Ink* is the perfect curriculum for all the helping professions: medicine, nursing, psychology, counseling, and coaching. It provides the science of neuroplasticity and expressive writing in a unique pairing that promises to create new positive tracks in the brain. The process allows one to embody a new understanding and gets the brain to maintain a new direction through self-directed brain change.

I found myself captivated as I engaged in the writing, and I wish the book were required reading for graduate students and residents. I am eager to use this in teaching and as adjunct homework for my clients, and I am grateful for a whole new understanding of brain neurology presented in simple and profound wording. Ross and Adams, your guides on the side, will open you up to new perspectives you didn't know how much you needed in order to be effective in your own life and those of others." —**Linda W. Peterson-St. Pierre**, PhD, MFT, MRN, emeritus professor, University of Nevada-Reno Medical School

"Adams and Ross are a wise and masterful team in the creation of this user-friendly trail map into the worlds of neuroscience and expressive writing. The Journal Ladder is clinical brilliance and indeed creates a sturdy bridge to enriched community of practitioners and researchers alike. As clinicians, it is important to be able to offer our clients ways to continue growth work between sessions. This book offers ten-to-twenty- minute writing 'bites' that are both safely and powerfully constructed.

Your Brain on Ink goes far beyond the goal of 'advancing the dialogue' about the role of expressive writing in healing. This is groundbreaking work." —**Nancy S. Scherlong**, LCSW, PTR-M/S, CJT, writer, adjunct faculty at Columbia and Adelphi Universities; expressive arts trauma therapist and positive psychology coach

"Ross and Adams have woven a unique workbook for self-awareness, mindfulness, healing, and growth, all in the context of modern brain science. Simple guided exercises are distilled from decades of scholarly research, making neuroplasticity available to all who seek self-directed change. Fresh writing and carefully constructed prompts invite readers to join the dance and observe their own brains in real time. This book provides fascinating and ready access to passion, clarity, and purpose." —**Nathan Ohren**, founder/director of Write4Life.com and host of the JournalTalk podcast

"*Your Brain on Ink* is a masterful dance between two expert and wise practitioners. In precise and conversational language, Ross and Adams explain and illustrate the fertile intersection between brain research and expressive writing. You are guided, step by small step, into a *sustainable* writing practice that will—literally—change the shape of your brain. From my viewpoint as a business coach and facilitator, this workbook is a quantum advance in how to help clients shift perspective and behavior by activating their own powers of expression and reflection to rewrite the brain maps they live and work by. This pioneering work will change many brains—and many lives—for the better." —**Edwina Cowdery**, business coach and facilitator, Sydney, Australia

"*Your Brain on Ink* is a treasure added to the expressive therapies field. This is the first therapeutic writing book that takes into consideration the neurobiological response to trauma and the sensory healing response it requires. Journal therapy is a natural medium to work with the amygdala's release process, and this book explains the manner in which this is accomplished and the profound potential healing effects. As the focus on trauma healing moves from a lens of trauma-informed care to trauma-applied practice, this book addresses both in practical, understandable, and applicable language. And because journal therapy is also a mindfulness-based practice, the astounding benefits of mindful brain changes are apparent and honored throughout. This is an exciting, succinct, and timely addition to the field."—**Cherie Spehar**, LCSW, CTC-S, RPT-S, founder/director/lead clinician, Apex Center for Trauma Healing, NC

"If we are as happy as we choose to be, to paraphrase Abe Lincoln, how can we become happier than we are, and feel that way more often? According to Ross and Adams, the answer comes from neuroscience: by frequently writing about experiences that leave us feeling more positive in the moment, we strengthen our brains' ability to do more of the same over time. Based on recent neuroscience discoveries and time-tested journaling techniques, this practice diminishes our brains' evolutionary mandate to focus on the negative, allowing us to observably shift our feelings toward the positive end of the scale in ten to twenty minutes of writing. And, like an athlete building muscles through repetitive movement, the more we use these techniques, the more our brains build and strengthen the neural pathways for positivity. In a world that seems to be drifting ever more negative, *Your Brain on Ink: A Workbook on Neuroplasticity and the Journal Ladder* provides a very real way to lift ourselves into more positive realms." —**Barbara Stahura**, CJF, coauthor of *After Brain Injury: Telling Your Story*

Editorial Review Board

It's Easy to W.R.I.T.E. Expressive Writing Series

Kathleen Adams MA, LPC
Series Editor
President/CEO, Center for Journal Therapy, Inc.
Denver, Colorado

Vanessa Furse Jackson, PhD
Poet, novelist, short story writer
Retired professor of English, Texas A&M University-Corpus Christi
Winchester, England

Peggy Osna Heller, PhD, LCSW, PTR-MM/S
Director, Wordsworth Center for Poetry Therapy Training
Registered poetry therapist and licensed clinical social worker
Potomac, Maryland

Linda Barnes, MS, CAPF-M/S, CJF
Director, FeatherStone Consulting
Retired college counselor, Rogue Community College
Medford, Oregon

Also in the It's Easy to W.R.I.T.E. Expressive Writing Series

Expressive Writing: Foundations of Practice
Edited by Kathleen Adams

Expressive Writing: Classroom and Community
Edited by Kathleen Adams

Expressive Writing: Counseling and Healthcare
Edited by Kate Thompson and Kathleen Adams

Writing with At-Risk Youth: The Pongo Teen Writing Method
Richard Gold

Women, Writing, and Prison: Writers, Scholars and Activists Speak Out
Edited by Tobi Jacobi and Ann Folwell Stanford

Transformative Language Arts in Action
Edited by Ruth Farmer and Caryn Mirriam-Goldberg

Through Students' Eyes: Writing and Photography for Success in School
Kristien Zenkov and James Harmon

The Teacher's Journal: A Workbook for Self-Discovery
Kathleen Adams and Marisé Barreiro

The Flourishing Principal: Strategies for Self-Renewal
Kathleen Adams and Rosemary Lohndorf

Your Brain on Ink

A Workbook on Neuroplasticity and the Journal Ladder

DEBORAH ROSS AND KATHLEEN ADAMS

ROWMAN & LITTLEFIELD
Lanham • Boulder • New York • London

Published by Rowman & Littlefield
A wholly owned subsidiary of The Rowman & Littlefield Publishing Group, Inc.
4501 Forbes Boulevard, Suite 200, Lanham, Maryland 20706
www.rowman.com

Unit A, Whitacre Mews, 26-34 Stannary Street, London SE11 4AB

Copyright © 2016 by Kathleen Adams and Deborah Ross

All rights reserved. No part of this book may be reproduced in any form or by any electronic or mechanical means, including information storage and retrieval systems, without written permission from the publisher, except by a reviewer who may quote passages in a review.

British Library Cataloguing in Publication Information Available

Library of Congress Cataloging-in-Publication Data

Names: Adams, Kathleen, 1951– author. | Ross-Swain, Deborah, author.
Title: Your brain on ink : a workbook on neuroplasticity and the journal ladder / by Kathleen Adams and Deborah Ross.
Description: Lanham : Rowman & Littlefield, [2016] | Series: It's easy to W.R.I.T.E. expressive writing | Includes bibliographical references.
Identifiers: LCCN 2016005003 (print) | LCCN 2016013530 (ebook) | ISBN 9781475814248 (cloth : alk. paper) | ISBN 9781475814262 (Electronic)
Subjects: LCSH: Diaries—Authorship—Psychological aspects. | Diaries—Therapeutic use. | Writing—Psychological aspects. | Creative writing—Therapeutic use. | Well-being. | Neuroplasticity. | Self-actualization (Psychology)—Problems, exercises, etc.
Classification: LCC PN4390 .A34 2016 (print) | LCC PN4390 (ebook) | DDC 615.8/515—dc23
LC record available at http://lccn.loc.gov/2016005003

∞™ The paper used in this publication meets the minimum requirements of American National Standard for Information Sciences—Permanence of Paper for Printed Library Materials, ANSI/NISO Z39.48-1992.

Printed in the United States of America

Contents

Series Overview: About the *It's Easy to W.R.I.T.E. Expressive Writing Series* ... ix

Foreword by Nicholas J. Mazza, PhD ... xiii

Preface ... xv

Introduction ... 1
    Deborah's Glossary and Guide to Your Brain ... 6
    Kay's Eight Suggestions for Satisfying Journal Writing ... 8

Section 1: A User-Friendly Guide to Your Brain and Your Journal

1. Prelude ... 13
2. Welcome to Neuroplasticity ... 17
3. The Art and Science of Expressive Writing ... 25
4. The Journal Ladder ... 35
5. Brain Maps ... 41
6. The Reflection Write ... 45

Section 2: The Write Way to Positive Brain Change

7	The Brain as Velcro and Teflon	51
8	Your Limbic System	57
9	The Brain Takes the Shape the Mind Rests Upon	61
10	Neural Darwinism	69
11	Neurons that Fire Together Wire Together	73
12	For Your Olfactory Delight	81
13	The Masking of a Negative Is Not a Positive	87
14	From Positive State to Positive Trait	93
15	Building Neural Circuitry	99
16	Series of Three	103
17	And Now for Something Completely Different	107
18	Anticipate a Blossoming of Creative Delight	117
19	Short Bursts of Radiance	129
20	A Radical Departure	137
21	Overwriting the Negative	143

Section 3: Our Last Collective Firings

22	Reprise: Your Brain Takes the Shape Your Mind Rests Upon	153
23	Bridging Into the Future	159

References	165
Literature Review: Evidence-Based Research on Expressive Writing	169
Acknowledgments	173
About the Authors	177

Series Overview: About the *It's Easy to W.R.I.T.E.* Expressive Writing Series

Expressive writing originates from the writer's lived experience—past, present, or imagined future life. Written in the author's own voice, expressive writing creates bridges between thought and feeling, reason and intuition, idea and action. It is equally rooted in language arts and social science, and it takes multiple forms: journals, poetry, life story, personal essay, creative nonfiction, song lyrics, notes, and snippets of thought. Expressive writing is democratic and accessible. No special knowledge is needed, supplies are available and affordable, and research confirms that outcomes can be profound and even life-changing.

The *It's Easy to W.R.I.T.E.* Expressive Writing Series captures the voices of worldwide experts on the power of writing for personal development, academic improvement, and lasting behavioral change. Authors are both theorists and practitioners of the work they document, bringing real-life examples of practical techniques and stories of actual outcomes.

Individually or as a compendium, the volumes in the *It's Easy to W.R.I.T.E.* Expressive Writing Series represent thoughtful, innovative, demonstrated approaches to the myriad ways life-based writing can shape both critical thinking and emotional intelligence. Books in the series are designed to have versatile appeal for classroom teachers and administrators, health and behavioral health professionals, graduate programs that prepare educators and counselors, facilitators of expressive writing, and individuals who themselves write expressively. Workbooks offer well-crafted, self-paced writing programs for individual users, with facilitation guides and curricula for anyone who wishes to organize peer-writing circles to explore the material in community.

Each book or chapter author is held to exacting standards set by the series editor, Kathleen Adams, who, prior to her 1985 launch as a pioneer and global expert in the expressive writing field, was trained as a journalist and served as chief editor for a nonfiction publishing company.

It's Easy to W.R.I.T.E.

*W*hat do you want to write about? Name it. Write it down. (If you don't know, try one of these: *What's going on? How do I feel? What's on my mind? What do I want? What's the most important thing to do? What's the best/worst thing right now?*)

*R*econnect with your center. Close your eyes. Take three deep breaths. Focus. Relax your body and mind. Gather your thoughts, feelings, questions, ideas.

*I*nvestigate your thoughts and feelings. Start writing and keep writing. Follow the pen/keyboard. If you get stuck, close your eyes and recenter yourself. Reread what you've already written and continue. Try not to edit as you go; that can come later, if at all.

*T*ime yourself. Write for five to twenty minutes or whatever time you choose. Set the timer on your phone, stove, or computer. Plan another three to five minutes at the end for reflection.

*E*xit smart. Reread what you've written and reflect on it in a sentence or two: *As I read this, I notice . . .* or *I'm aware of . . .* or *I feel . . .* Note any action steps you might take or any prompts you might use for additional writes.

Foreword

Kathleen Adams, a licensed professional counselor and internationally recognized leader in journal therapy, poetry therapy, and therapeutic writing, teamed up with Deborah Ross, a psychotherapist and journal therapist with expertise in neuroplasticity and mindfulness. Their collaboration resulted in this tenth and final book in the *It's Easy to W.R.I.T.E. Expressive Writing Series* for which Kay is series editor. Given Kay's prolific and profound contribution to the professional literature on therapeutic writing, it is astounding to see that she and Deborah offer still another innovative and compelling work on expressive writing that is based on science, the social sciences, and literary perspectives.

The authors' focus on the intersection of expressive writing and neuroscience is consistent with the expectation of evidence-based practice. Deborah captured the essence of the central theme of the art and science of expressive writing for health and mental health purposes: "Science is now able to validate what poets, philosophers, and positive thinkers have intuited as far back as Aristotle: *We can change our brains by changing our minds. We can change our minds through the power of our intention, attention, and self-directed action*" (p. 41).

Kay's vision of expressive writing as a part of a holistic approach to advance health, mental health, and personal growth is realized in this book. The range of human experience (cognition, affect, behavior, and spirit) is sensitively and perceptively addressed through the provision of overviews of the supporting literature, guides, exercises, and examples. Within Kay's Journal Ladder (moving from highly structured to unstructured), a three-stage process for expressive writing is presented. This includes *intention* (starting), *attention* (the writing itself; exploring thoughts, feelings, etc.), and *action* (reading back and reflecting).

Due to the amazing ability of the coauthors to present in conversational style with pauses for constructive action for the readers, *Your Brain on Ink* reads like a great story that inspires and activates the reader. This book is much more than a workbook; it is poetic validation, inspiration, and engagement. Like all good poetry, it is unfinished. Each time a reader responds to a poem, a new poem emerges. Each time you read this book (best taken in segments), a new chapter or verse will emerge that has the potential to change your life.

Consistent with cognitive-behavioral, narrative, and positive psychology, the authors draw together established writing exercises (e.g., clustering, alphapoems, imagery, sensory prompts, and sentence stems) with their own established journal methods and a consistent directive to provide and explore three feeling words before and after each write. These exercises and the overall format/content of the book engage the reader to consider cognitive restructuring and behavioral actions, writing their own "story," and focusing on their strengths. This also leads to community engagement and growth.

I am thankful to the authors for not only the privilege of writing this foreword but also the opportunity to benefit from "testing out" the exercises on myself. The book is very sound, user-friendly, and appropriate for a wide range of readers who have different levels and types of professional experience. It is an indispensable resource for those interested in the study and ethical practice of expressive writing in educational, therapeutic, growth, and community-building activities.

Nicholas F. Mazza, PhD
Dean and Patricia V. Vance Professor Emeritus
College of Social Work, Florida State University
Editor, *Journal of Poetry Therapy*
FL Licensed Clinical Social Worker, Psychologist, Marriage and Family Therapist
Author of *Poetry Therapy: Theory and Practice, 2nd Edition* (2016, Routledge)
nfmazza@fsu.edu

Preface

There is nothing so wise as a circle.

—*Ranier Maria Rilke*

KATHLEEN ADAMS

Renku is both a poetry parlor game and a community experience. Like haiku, it is a Japanese form dating back many centuries. A *renku* is a multi-stanza poem (often thirty-six verses) passed from hand to hand at a gathering of friends, family, or poets. Each person in turn writes a new stanza, either a three-line haiku or a two-line variation; the *renku* makes the rounds several times. There are conventions for the construction, but the most important feature of *renku* is that each verse must *link* to the verse preceding it; this then leads to a *shift* into something new. This linking and shifting, connecting and extending, coming together and breaking apart informed by what came before, is at the heart of *renku*.

Dr. Hiroshi Tamura, a Tokyo psychiatrist, is a *renku* master. He uses it in his work with schizophrenic patients (Tamura 2001). When Hiroshi comes from Japan to the annual poetry therapy conference every few years, someone always hosts a *renku* party. Hiroshi serves as our *sabaki*, our facilitator.

Writing this book has been like writing a *renku*. My coauthor Deborah Ross and I have linked and shifted, touched back and moved ahead, constantly merging and then again separating the two fields, neuroplasticity and expressive writing.

Deborah writes the next stanza:

> From the world of poetry and *renku* we come to the world of science and the phrase *differentiation and linkage* (Siegel 2012). Kay and I each brought years of training and engagement with questions about how to facilitate healing, growth, vitality, and resilience in ourselves as individuals and in the populations and communities that we serve. This book represents the neurobiology of "we": the ways that we toggled back and forth linking ideas, uncoupling them, recoupling them, all in the service of building a third entity, this workbook. We were guided by the wisdom of our respective trainings, by honoring the integrity of the disciplines, by maintaining their differentiation as we linked them, and by our deep respect and caring for each other. We hope that you, the reader, can use our *renku* as a guide to looking inward with your pen in service of your own healing, well-being, vitality and integration. Your linkages with others will undoubtedly be richer for it. (Deborah Ross)

The Journal Ladder

I met Deborah in 2010 when she attended my one-day continuing education seminar for psychotherapists on the Journal Ladder. The Journal Ladder is my signature expressive writing method that lays out a clear, easily implementable model for offering journal writing to individuals at all levels of experience and with all manner of presenting problems. In the last decade I have taught this class to more than 8,000 therapists across the United States, and nearly all of them tell me that the Journal Ladder completely shifted the way they think about the use of writing in therapy.

The Journal Ladder was originally designed as a trauma model, one developed while I was working as a journal therapist for a psychiatric hospital program specializing in dissociative disorders (Adams 1998). Most of the patients were already writing journals on admission, and they reported positive relationships with writing, although nearly everyone also told me harrowing stories of "falling off the cliff" into deep depression, anxiety, or traumatic reenactment because writing had triggered difficult memories or otherwise overstimulated them.

The Journal Ladder created a continuum of writing techniques, moving from the most highly structured, paced, and contained at the bottom of the ladder to the most unboundaried and unstructured at the very top of the ladder—the "cliff" the patients kept falling off. It taught patients how to self-regulate their written disclosures by shifting to a technique in the lower rungs of the ladder, or otherwise intentionally adding structure, pacing, or containment, when they began to feel out of control while writing.

Later, when I started a private therapy practice, I realized that the Journal Ladder was a solid clinical model that was also useful in helping clients with all sorts of challenges—both life challenges and writing challenges—find manageable entry into the benefits of expressive writing.

Now the Journal Ladder has come full circle: it grounds and anchors the actualization of human potential through positive brain change. The combination of the Journal Ladder with neuroplasticity has given rise to what may well be a revolution in expressive writing: it offers articulation, documentation, and visibility in similar ways that the first functional MRIs allowed neuroscientists to actually look into a working brain.

Self-Directed Neuroplasticity

In the preface to *Expressive Writing: Counseling and Healthcare* (2015), the eighth book in this series, I wrote:

> A recent focus (and perhaps the next big shift) [in the modern history of expressive writing] is the relationship between expressive writing and what Deborah Ross (Chapter 2) calls "self-directed neuroplasticity." Several other chapter authors also share their thinking about the intersection of writing and neuroscience. The emerging work will take up the study of how a purposeful and intentional program of expressive writing designed specifically around self-directed or practitioner-guided goals for healing, growth, and change can create neural pathways that over time may literally change the brain. (Adams 2015, xviii)

In this, the tenth and final book in the series, Deborah and I extend her thinking about using expressive writing for self-directed neuroplasticity into a workbook of applied practice. The workbook offers a three-stage writing process for self-directed neuroplasticity: the *intention* stage is the act of picking up the pen and invoking the power of writing; the *attention* stage is the writing itself, the focus on and exploration of one's own thoughts, feelings, beliefs, memories, experiences, and desires; and the *action* stage is the reading back of what was written and harvesting its insights, questions, or solutions through the follow-on reflection write.

These stages—*intention, attention, action*—represent a working model for strengthening or creating neural circuitry, as well. Here is Deborah telling us why:

> One of the few times in my life when I am not multitasking is when I am writing in my journal.
>
> Because I have an established journal practice, all I have to do is pick up my journal, and the circuitry that has been created around that practice starts to fire. Then I sit with my journal and have a real-time experience with writing, during which I may well be strengthening an existing circuit that enhances my life, or I could be wiring something new. Although I am not focused on creating flow, that is a natural byproduct.
>
> My journal is asked to be the witness to my experience, and then I witness my journal. Taking the action of the reflection write usually reveals an invitation to a next step: a new experience, an action in service of a goal, a different way of thinking or feeling, a broadening of a horizon. Whether these steps represent action that is internal or external, neurons are firing in association with it. Over time and with practice, these neurons wire together. (Deborah Ross)

What You Will Learn

In this workbook you will learn interweaving tracks on the neuroscience of brain change and the journal techniques, tools, and writing prompts that will help you write your way into the more positive state you desire. Additionally, you will learn success secrets such as:

- Why our brains are hard-wired to be "Velcro for the negative and Teflon for the positive" (Hanson 2011)

- How to observably shift your feelings in under twenty minutes
- Why your nose is a tremendous ally in positive brain change, and how to use it for your good
- How to search out and rewrite the masked negatives in your speech and writing
- Why novelty and creativity are superfoods for brain change
- How the brain and writing both engage in map-making
- How to overwrite the negative and bring healthy positive associations to difficult experiences
- Why short bursts of radiance can accelerate the creation of new neural circuitry

Who This Book Is For

The audiences for this book include: counselors and therapists; neuroscientists and those who practice methods based in neuroplasticity; health care professionals; practitioners and facilitators; and writers who know or are discovering the power of expressive writing for healing, growth and change.

If you are a counselor or therapist, you will have a handbook on two of the hottest topics in psychotherapy today: the emerging science of neuroplasticity, and the role of the personal process journal in the therapeutic relationship.

If you are a neuroscientist, or a student or practitioner of neuroplasticity methods, you will learn the power of adding expressive writing to accelerate behavioral and emotional healing, growth, and change.

If you are a health care professional, you will discover new pathways to help patients find holistic self-management strategies for real-time relief.

If you are a credentialed practitioner, community-based facilitator, writing coach, creative writing teacher, or other who brings this work to the world, you have guidance and role-modeling from master teachers who can inform and bring insight to your work.

If you are a writer or seeker of personal growth and development, you hold in your hands a method that represents an exciting new frontier in self-help. Science tells us we can change our brains to embrace more resilience, peace of mind, and fulfillment, and this workbook demonstrates how.

"There Is Nothing So Wise as a Circle"

In 1985 I called my first journal circle. Six of my friends sat on my living room floor and wrote their hearts out and shared their stories. I was a first-semester graduate student in a counseling program, and I knew in that instant that the intersection of personal healing and journal writing was my life's work.

I have always done my work in community. I am a natural weaver of lives. I am a gatherer of overlapping, intersecting circles of those who speak on the page, who read their stories in voices that quaver or whisper or soar, in the presence of witnesses who receive without judgment and respond with compassion.

This is what I have learned in writing circles across the decades and around the world: Expressive writing changes us. It brings peace to our hearts. It restores balance. Writing connects us to the parts of ourselves that have been lost, abandoned, neglected, forgotten, or ignored. We learn that there are wise, friendly, and reasonable aspects of ourselves that want us to succeed. These "best selves" are willing to share their wisdom with us. Writing is the bridge.

Community also changes us. We learn that we are not alone. We see ourselves reflected in another's compassionate eyes. We hold for each other what we cannot reliably hang onto for ourselves. We learn that we are enough, we are sufficient, we have something to say and a voice with which to say it.

Pennebaker states that "expressive writing has the potential to change the way you see the world" (2013, xiv). I believe that writing communities—in classrooms, conference rooms, family rooms, therapists' offices, places of worship, community centers, coffee shops—have the power to deepen and accelerate the process of change. What might arise, I wonder, from a collectively shifted world view?

My vision for this book, (and indeed for the entire ten-book series, which with this volume comes to closure) is that it will substantially advance the dialogue about the role of expressive writing for healing, growth and change at every level of body, mind, and spirit: in the mirror, at the dinner table, in the therapy chair, at the doctor's office, in the classroom, at the place of worship, throughout the community, into the culture.

I welcome you to the community formed by those who read this book, and I invite you to join the conversation.

Kathleen Adams MA, LPC
Series Editor, *It's Easy to W.R.I.T.E.* Expressive Writing Series
May 2016

Join the conversation:
www.itseasytowrite.com
kathleen@itseasytowrite.com

Introduction

KAY ADAMS

Maybe you've always heard that writing a journal would be good for you. Maybe you've been writing a journal forever and swear it's saved your life, marriage, or career. Perhaps you're a start-and-stop journal writer who has had mixed results over time, or your journal never quite kicked into gear. Maybe you're fascinated with the idea of neuroplasticity and want to know more about how to change your brain for a more positive, productive, purposeful life.

I'm Kay Adams, psychotherapist and director of the Center for Journal Therapy in Denver, Colorado. My "tall" name is Kathleen, but I go by Kay in person. For more than thirty years my work has taken me to the intersection of therapeutic writing and self-actualization. I create, study, and teach ways to use journal writing for healing, growth, and change.

One of the great joys of teaching is encountering students who stand on the shoulders of the learning and create fresh new avenues of theory and practice. I first worked one-on-one with Deborah during her studies in journal therapy at the Therapeutic Writing Institute, the on-line professional training program I founded. In my curriculum design class Deborah articulated and actualized her vision for a program called *Your Brain on Ink*. I remember individual moments of recognition that Deborah was about to make history for being one of the very first to combine informed writing practice and neuroplasticity.

Shortly thereafter we began teaching together. Now, for the first time, this workbook articulates our findings on the power and potential of positive brain change, using crafted writing processes to support and reinforce the foundational principles of neuroplasticity. You are about to write your way into your own possibilities—the qualities that will help you live a fuller, richer, more authentic life.

Deborah Ross

As a psychotherapist in private practice for many years, I've been a student of those models of therapy that are relational. It doesn't matter to me if it's the relationship that we have with ourselves and the myriad parts that make up our internal family, or the relationships that we engage in with others, or the relationships that we have with our communities and the world at large—if it's about relationships, I want to know about it.

In 2011 the axis of my world tilted in a life-altering orientation as my personal relationships with journaling, neuroplasticity, and meditation came together in one big braid. I was studying journal therapy at Kay Adams's Therapeutic Writing Institute and interpersonal neurobiology at Dan Siegel's Mindsight Institute. I was maintaining a mindfulness meditation practice, and I was regularly writing in a journal.

I could feel connections and cross-fertilizations. I could grasp a greater sense of the workings and power of my mind and brain. That knowledge met on the pages of my journal and coalesced in my studies of therapeutic writing, yielding an understanding of how I could more intentionally, consciously, purposefully partner with my pen to author greater vitality, resilience, healing, and joy in my life.

There is a "circuits afire" *knowing* that comes sometimes. It is the magic moment of synthesis, and I had one of those *knowings* as I held the deeply beautiful, powerful relationship between the ancient path of writing, as modernized for healing and positive change by 20th- and 21st-century thinkers, and the cutting-edge path of neuroplasticity. I invite you to join us as we integrate the two, writing and neuroscience, to help you write the story, whatever it may be, that you truly want to be living.

Kay Adams

This book is a conversation. It's a conversation between Deborah and me, between us and you, and among participants in Deborah's classes. Your voice joins the circle each time you write and reflect.

In order to keep all these conversations straight, we've organized the chapters like this:

Teachings: Most chapters start with a brief "teach" from Deborah on the neuroscience followed by one from me on the writing. Sometimes I start and Deborah follows, but nearly always I have the last word, because I introduce the Inklings.

Inklings: These are your writing prompts, with instructions.

Reflections: When you complete a write (an Inkling), you'll read it and give yourself process notes about the write itself, how it landed in your body, what you noticed, or where you might take action. This is your opportunity to harvest insight from your writing.

Conversations: Many chapters contain replications of conversations that originated between Deborah and students in her online classes.

Three feeling words: At the start of the first Inkling and the end of the last Reflection in any chapter, you'll see the title *Three Feeling Words* and three lines. When you see this cue, close your eyes and check in with yourself. Find three feeling words or phrases that describe your current state, and jot them down. Then compare your befores and afters.

This small convention offers big paybacks in observable data about the way your feelings shift with time and process, the *time* usually being under twenty minutes and the *process* the Brain on Ink writes and reflections. You may soon recognize that you are likely to feel differently (and often better) after writing and reflecting than you did just minutes before.

Now let's gather your gear. You won't need much, just a pen or pencil you like to write with and a fresh new journal (as simple as a spiral notebook or fancy as a blank bound book). If you're opting for a keyboard, create and name a digital file My Brain on Ink. And bring along this book, of course; it will be your guide and resource.

If you already write a personal process journal that holds the entire spectrum of life experience, including the oscillations of mood, temperament, satisfaction, and the inner and outer explorations of life as you're living it, by all means bring it, too. However, take care to keep it separate from the Brain on Ink journal.

We want you to start a new notebook or file for the Brain on Ink journal because it will be important for you to be able to see the progression of your writing, reflecting, and thinking in real time. Date every entry to create a timeline.

If you're not already writing a journal, that's about to change. You'll be writing a specialized journal, not one that will hold your entire life, with its day-to-day successes, flops, and accompanying feelings. You're welcome to write that journal too, if you choose. If you do, keep it in a different notebook or file. The Brain on Ink journal is just for the prompts and reflections in this workbook and any observations you may make about the process.

Deborah Ross and Kay Adams

In the world of neuroplasticity, it is known that certain components add to the capacity of the brain to change itself, for better or worse (we're focusing on "for better"). These include intentionality, focus, mindfulness, relationships, reinforcement of the positive, and a strong wash of pleasant or welcome emotions such as love, joy, enthusiasm, lightness, or connection.

We believe three components that particularly shape the attainment of positive brain change through expressive writing are *intention, attention,* and *action*. These components are replicated time after time in the expressive writing process:

- The *intention* stage begins when we pick up a pen or open a file to write in a purposeful way. The journal devices for this stage are the prompt and the entrance meditation, which each set the stage for the question, *What do I want to create?* The quality for the *intention* stage is the creative spark; neurologically, the task is to fire up neurons, make synaptic connections, and create or strengthen neural circuitry.

- The *attention* stage is next. Having set intention, we begin observing with a curious and compassionate gaze. The journal device is the write itself, which asks the question, *What is the story I am telling?* The quality for the *attention* stage is focus; neurologically, the task is integration using both hemispheres of the brain—the specific integration being aligned with the intention.
- Finally we come to the *action* stage. Having cultivated focus through writing, we shift into movement, the *doing* after the *being*. The journal device is the reflection write, which asks the question, *What do I cultivate and harvest?* The quality for the *action* stage is embodiment; the neurological task is to support the conscious mind in everything from making that one phone call or finishing that one project to living ever more intentionally, with meaning, compassion, and purpose.

On the next pages you'll find two resources, Deborah's Glossary and Guide to Your Brain and Kay's Eight Suggestions for Satisfying Journal Writing. Take a glance at them now, know they're here for reference, and let's get started!

Deborah's Glossary and Guide to Your Brain

Amygdala: The brain's alarm bell. Centrally located in the limbic region, it's involved in the activation of emotion, processing of social signals, and the appraisal of meaning.
Attention: The process that regulates the flow of information. We can aim it, sustain it or shift it.
Axon: See Neuron
Brain: Often refers to the skull or "head brain," but can mean the extended nervous system in the body that includes the neural processors around the heart and gut.
Dendrite: See Neuron
Hippocampus: Centrally located in the limbic region, it plays a major role in memory and gives the brain a sense of the self in time and space.
Insula: Located in the middle prefrontal cortex of the brain, it links information from the body to the higher cortical regions of the brain. Its link to other structures in this area is associated with self-awareness.
Limbic system: Sometimes referred to as the emotional regulatory center, it evolved during the period of mammalian evolution and includes the amygdala and hippocampus. Coordinating signals from the body, brain stem, and higher brain regions, it is also responsible for some key mental processes involving the triggering of emotion, the processing of social signals and a determination of meaning.
Memory: Explicit memory requires focused attention and is factual/autobiographic: recalling the bike ride that you took last week. Implicit memory does not require focused attention and can include perceptions, emotions, body sensations: riding the bike itself.
Name it to tame it (Siegel 2012): Placing a label on an experience can calm the mind and stabilize attention.
Neuron: A basic cell in the nervous system that has three parts: dendrites, cell body and axon. Dendrites are short, branched, tree-like projections that receive information from other neurons. They lead into the cell body which is responsible for maintaining the cell and contains its DNA. Leaving the cell

body is the axon, a single strand insulated cable that carries the electrical impulses to the waiting dendrites of the next cell.

Neurons that fire together wire together: Donald Hebb's (1949) proposal and Carla Shatz's articulation (1992) that neurons that happened to fire together once will likely fire together in the future because synaptic linkages were created during the first firing. Freud proposed essentially the same thing in his Law of Association when he said that events experienced together will be associated with each other in memory (Amacher, 1965).

Neural Darwinism: "Use it or lose it; survival of the busiest." Neural connections that aren't used are pruned or wither away. It was first described by Gerald Edelman (1978).

Neuroplasticity: The ability of the brain to change itself in response to attention and experience. How you use your mind or focus your attention changes your brain.

Parasympathetic nervous system: One of two branches of the autonomic nervous system; it is the quieting branch and calms heart rate, respiration and alertness. The parasympahetic nervous system is associated with the vagal nerve. It has a braking action on the system. It can be activated with deep or calming breathing.

Self-Directed Neuroplasticity: Our ability to direct brain change, either positively or negatively, by how we consciously focus our attention. Repeatedly nurturing a list of gratitudes creates different brain circuitry than repeatedly feeding a list of grievances. As used in this book, a method of accelerating experience-dependent neurological change through expressive writing.

Kay's Eight Suggestions for Satisfying Journal Writing[1]

Since there are no "rules" for expressive writing (for instance, you can relax about correct spelling or grammar or whether you're making sense—and you can write by hand or keyboard, as you wish), it's easy to succeed! However, many people benefit from keeping these suggestions in mind.

1. **Be mindful of privacy.** Discreetly stash your journal in its own special place (bookbag, night stand, desk drawer) for peace of mind and protection from curious eyes. If you're writing on a computer, give your files neutral names or password-protect them.
2. **Start by centering.** Expressive writing benefits from turning inward. Before you write, close your eyes, take a full deep breath, and let go of tension in your body and mind. Repeat a few more times. Let your mind gather itself around the area you'd like to explore.
3. **Date every entry.** If there were a "rule," it would probably be this one. Dating every entry allows you to construct a chronological narrative. It also lets you hear the silence between your entries.
4. **Read and reflect.** Read each entry and give yourself a sentence or two of reflection: As I read this, I am aware of— or I notice that— or I'm surprised by— or I feel/think/want—. This reflection synthesizes expression and intuitive understanding.
5. **Write quickly.** You can outsmart writer's block by writing so fast that the Inner Critic and the Inner Editor can't keep pace.
6. **Start writing; keep writing.** Once you begin to write, keep the pen moving. You can edit and fix glitches later, if you want, although it isn't necessary.
7. **Tell the truth as you know it.** Writing it down doesn't make it "real." It does make it noted and acknowledged. When you tell the truth as you know and experience it, you make room to challenge your assumptions, acknowledge your current reality, release secrecy, and change your mind/heart/behaviors.

8. **Write naturally. Do what works.** If you're more comfortable keyboarding than writing by hand, that's fine. If you want to write in bullet points or poetry, do it. If you change your mind about something you wrote, tear the page out or make a collage over it. Stay open to your own inner wisdom and intuition, and follow your own internal guidance.

Note

1. Adapted from Adams 1990, 27–43.

I

A USER-FRIENDLY GUIDE TO YOUR BRAIN AND YOUR JOURNAL

1

Prelude

Deborah Ross

I first began to keep a journal as a teenager. It was one of those locked Dear Diary volumes that received the outpourings of my heart as well as the more mundane details of my life. But someone read it, and I stopped writing.

Years later, after a series of compounding losses, I found myself pouring my heart out again, this time to a therapist. We used my own life story along with fairy tales, Zen koans, song lyrics, and poetry in the service of constructing a healing narrative. I didn't author the words, but I surely learned the power of stories.

About this time, research on the benefits of aerobic exercise was capturing headlines and attention. One of the most interesting discoveries was the evidence that cardiovascular workouts stimulated the release of endorphins, the "feel good" neuropeptides responsible for the legendary "runner's high." My therapist suggested that I start running, hypothesizing that if my brain got a regular natural dose of feeling good, it might help lift my depression.

So I started logging miles. Journaling came back into my life with a notebook I used as a runner's log. In it I recorded routes, shoes, mileage, adventures with my dog, and the occasional observation of an ease that was creeping into my life.

After twenty years of running, logging, and experiencing ease, I signed up for the Intensive Journal® workshop created by one of the most influential minds in expressive and therapeutic writing, Dr. Ira Progoff (1992). The words on the page felt like those early steps on the trail. The personal journal was making its way back into my life.

It was many more years before I could say that *journal* was both a verb and a noun, both a practice and a product. But it was becoming as much a part of me as my running (now walking) practice.

I noticed that I was not just recording and reflecting about the current unfolding of my life. I was also using my journal to integrate years of learning, experience, observation, and celebration:

- I most fully transitioned to being an athlete after I wrote about my first race, years prior.
- The embodiment of putting one foot in front of the other was on the page, ready to be claimed as a guidepost and as documentation of incremental success.
- I explored *marathon* as both event and metaphor. The exploration informed the way I cultivated my own resilience and self-care strategies.
- Snippets of poetry and those Zen stories that I'd heard in therapy long ago became subjects of writing and reflection. These writes became the inspiration for my mindfulness practice.
- An ongoing integration felt like a deep well that quenched my soul's thirst.

Just as I had developed running/walking routes, so too had I developed brain circuitry that routed my dominant experience toward resilience, growth, and determination. Today I still walk, I still write, I still practice mindfulness, and my brain fully engages with it all. I live an embodied transformation.

Kay Adams

In the spring of 1985, five months before I started graduate school in counseling and six months before I created and began teaching *Write On!* (now *Journal to the Self*) workshops, I entered a phase of intense personal growth. I felt creatively on fire. My learning curve was a scary, exciting roller coaster. I was living out loud, guided by intuition and core values. It was a prolific, yeasty time.

In only a handful of months, I noticed that I was integrating a different outlook on life. I was more positive, optimistic, grateful, and self-confident. I became more skillful with risk-taking, and I began to push the limits of my comfort zone. Five years later I barely recognized myself or my life.

I still have the journals from those years, detailed play-by-plays of my earliest beginnings as a teacher, author, and therapist. My journals document a process that makes my self-directed change observable—from the daily self-talk to the premonitions and foreshadows of work that would not emerge for years, even decades.

Would those changes have been permanent had I not written them down? I don't know. I do know that more than thirty years later, my work has continued to grow, and I have continued to take risks, create outcomes, and contribute to knowledge of lasting value. Throughout it all, I have used my journal as a living laboratory. Every good idea I've ever brought to the real world—books, courses, theories, conferences, the Journal Ladder, an entire school—was incubated in the pages of my journal.

I now recognize that long-ago phase as an intense time of *self-directed neuroplasticity*.[1] Decades before I knew its name, I was changing my brain by changing my mind. There was a synchrony of mental, emotional, spiritual, and behavioral *intention, attention,* and *action,* and they alchemized into what I now recognize as *new neural pathways* that actually shifted my world view and the preponderance of my experience. It was a thrilling time that yielded a new and exciting path of flow, one that my energy, emotions, thoughts, behaviors, and actions have followed since.

Now it's your turn. Has there been a time in your life when you have changed some aspect of your life through intention, attention, and action? Your first writing process—we call them "Inklings"—is next. You'll get an orientation to journal writing in an upcoming chapter; for now, just know that you can tell your story in whatever way you wish. Use the instructions as guidelines. Bring your *intention* and *attention* to the prompt, and let the *action* be putting pen to paper or fingers to keys to discover what emerges.

Inklings

My Story

Instructions: Has there been a time in your past or present when you have changed your life through the power of your intention, attention, and action? Set a timer for ten minutes, and write the story of your own self-directed change. If you wish to write longer than ten minutes, you are welcome to do so.

Three Feeling Words

_____

_____

_____

Note

1. Deborah's Glossary and Guide to Your Brain in the Introduction defines this and other neuropsychology terms.

2

Welcome to Neuroplasticity

DEBORAH ROSS

Welcome, welcome, welcome! The brain loves repetition. Kay and I invite you into a grand experiment. This workbook offers a multitude of ways to use expressive writing for the development of new neural pathways that can result in increased well-being. As you learn and apply the principles of neuroplasticity to your own life, staged and sequenced through masterful application of writing techniques designed to promote positive brain change, you may note improved satisfaction in various domains.

Neuroplasticity is the ability of the brain to change its structure in response to experience. That experience is often a mental process such as perceiving, thinking, feeling, or remembering. It may be a process of focusing, sustaining, and shifting attention. It may be emotional, sensory, or imaginal in nature. The experience can be, and often is, a multimodal combination of two or more processes.

The science of neuroplasticity demonstrates that your brain can and does change, through the activities of neurons, the basic cells of the nervous system, which, when activated in particular patterns, can grow stronger, fire more efficiently and begin to create new neural pathways in the brain. As individuals, we each not only have the power but use the power of synchronized firings to create neural pathways—for better or for worse. Unfortunately, left to our own devices, we will often create pathways that become ruts of stuckness, pessimism, even depression or anxiety. But here's the great news: We can also create pathways that lead to resilience, vitality, greater peace of mind, and improved well-being.

Experience-Dependent and Self-Directed Neuroplasticity

Rick Hanson (2011) suggests a modification to the term neuroplasticity. He calls it experience-dependent neuroplasticity, referencing the actual way the brain can change through accumulated experience.

Experience-dependent connections are created by the very experiences that you have as your life unfolds. Your individual world is full of sight, sound, scent, taste, and touch, each of which comes with particular and unique associations and meanings: the food served in your home, the sounds accompanying meals, and the people with whom you share your meals. These experiences activate neural firings that create synaptic connections unique to the meaning you create.

The brain's capacity to change is largely based on how we direct our attention. We can consciously focus on mindfulness—yoga, breathing, meditation—and the brain will create a new path of least resistance in that direction, bringing with it greater flexibility, adaptability, and ease. The brain, however, is predisposed to expect disaster as a mechanism for survival (*is that a snake or a stick?*—our limbic systems still follow the preservationist instincts of our earliest ancestors), and so it is actually physiologically easier to get and stay in ruts of rigidity, brittleness, or struggle.

My coauthor, Kay Adams, started the Center for Journal Therapy in 1988 with the mission statement to "make the healing art and science of [expressive] writing accessible to all who desire *self-directed change*" (Adams 2006a) (italics added). The term self-directed neuroplasticity was first used by psychiatrist Jeffrey Schwartz to reference the process of using the power of conscious focus to build new neural circuitry that supports healthier, more resilient thinking, doing, and being (Schwartz and Begley 2002).

In the *Your Brain on Ink* lexicon, self-directed neuroplasticity is a method of accelerating experience-dependent neurological change through expressive writing. It's what happens when you harness the power of your pen or keyboard in service of intention, attention, and action to stimulate healing, vitality, resilience, and growth. That's what we'll be working with in this book.

Kay Adams

Here's how the writing part of this book will look. In my part of the conversation, I'll tell you about the journal technique we'll be using, its form and function, and why it's a good choice for self-directed neuroplasticity. I'll also offer a preview for the writing to come.

Nearly everyone we work with prefers a handwritten or keyboarded journal to writing in an actual workbook, so we have given you a symbolic few lines at the end of each Inkling to indicate writing space.

Most Inklings conclude with an opportunity to read and harvest insight. We call this the Reflection, and you'll learn more about it soon. In some chapters this is followed by a section called Conversations; these are excerpts from dialogues between students and Deborah in a class at the online Therapeutic Writing Institute or from clients from our private journal therapy practices.[1] You'll read actual journal entries or reflections and Deborah's (and sometimes my) feedback, observations, and pointers. Some chapters end with standalone material tied in with the theme of the chapter.

Now let's continue with your own writing processes. At the end of this section, you'll find a set of five sentence stems. The sentence stem technique is the first rung of the Journal Ladder (a concept you'll learn about in the next chapter) because it maximizes structure, pacing, and containment, three vital journal tools. The sentence stem technique works for just about everybody; it's quick and easy.

When several sentence stems are compiled into one template in which each informs the other, we call it a structured write, and that's what you'll be experiencing.

Write whatever comes in response to the stems. Your writing doesn't need to look any particular way; all ways are equally "write."

The purposes of this Inkling are to dip your figurative toe in the ink and to create a baseline of thoughts that we'll return to down the road for reality and progress checks. It will take five to ten minutes to complete the sentence stems—five to seven if you're a spontaneous thinker, or seven to ten if your style is to allow thoughts to first gather and then bubble to the surface.

When you have completed the structured write, turn to the Reflection page and take another two or three minutes to read and give yourself a few sentences of reflective feedback on what you noticed in your writing.

That's it! That's your brain on ink.

Inklings

Self-Directed Neuroplasticity: Your Brain on Ink

Instructions: Complete each of these sentence stems. Be as specific as you can.

Three Feeling Words
_____
_____
_____

When I think about changing my brain, I notice . . .

If I could transform negative self-talk, behaviors, attitudes, and expectations, I would start with . . .

If I could create new pathways to healing, vitality, resilience, and/or growth, I would first want to create . . .

My vision for myself one year from today is . . .

My vision for myself five years from today is . . .

Reflection

Instructions: Read your sentence stems. Then give yourself a sentence or two of reflective feedback. You can start with either of the prompts below.

- As I read this, I notice (or: I am aware of, I see, I'm curious about, I'm interested in, I think, I feel, etc.)
- Where in my body do these responses "land"? Where do I feel this write? Note any somatic experience.

Three Feeling Words

_____
_____
_____

Conversations

Michael, a psychotherapist and student at the Therapeutic Writing Institute, offers thoughts on his own self-directed neuroplasticity; Deborah follows with response.

Michael writes:

When I think about changing my brain, I feel . . . hopeful, empowered, freed, liberated, excited and curious.

If I could transform negative self-talk, behaviors, attitudes and expectations, I would start with . . . addressing the thoughts, "sometimes I just don't measure up, sometimes I just don't get things right, sometimes I feel like I won't be appreciated;" responding in new ways with the mantra, "I have all that I need within me."

If I could create new pathways to healing, vitality, resilience and/or growth, I would first want to create . . . . a new neural-emotional-spiritual-relational highway that connects the 8 Cs of Internal Family Systems Therapy: compassion, clarity, curiosity, calmness, courage, creativity, connectedness and confidence.

My vision for myself one year from today is . . . celebrating and sharing within a wider community rather than going it alone.

My vision for myself five years from today is . . . living a more centered, joyful life, focusing on giving rather than receiving, letting go rather than holding on, emptying myself rather than filling, cooperating rather than competing.

Deborah responds: I'm a practitioner of Richard Schwartz' Internal Family Systems as well, so I'm happy to see you referencing it. The 8 Cs of IFS writ large and connected with your pen are certainly great circuits to fire in service of leaning into visions!

Something to be aware of and track in your reflection to notice writing process and language is how often the negative shows up as *don't, won't, couldn't, shouldn't,* etc. It's the language of constriction. And when you are creating a vision, such as the five-year one, you might try reading it out loud, first exactly as you wrote it and then a second time saying only the positives, allowing them to flow and link. You might notice a difference in your body with how you experience those two readings. I'm guessing that the second one, in which the positives link without the reference to what you no longer want, might feel more consistent with the hope and liberation you expressed in the first sentence stem.

Note

1. All examples are used with permission. TWI students chose their own *noms de plume*. Clients are composited and names have been assigned.

3
The Art and Science of Expressive Writing

KAY ADAMS

When I was ten years old I discovered the magic of writing down my life. I started with the little locking five-year diary, a feminine "write of passage" in the 1960s. In junior high I graduated to a Big Chief tablet and a purple ballpoint pen. Over the years, I have filled hundreds, maybe thousands, of notebooks and journals.

Writing down my thoughts, feelings, questions, concerns, dreams, confessions, and inspirations has been a major contributor to the development of my emotional intelligence, problem-solving abilities, and healthy self-concept. My journals and notebooks are curious, receptive friends with tireless capacity to mirror truth and understanding.

The research that psychologist James W. Pennebaker and his colleagues have gathered since the mid-1980s demonstrates that expressive writing helps improve both physiological and psychological well-being. There are various theories as to how and why this happens, but in general there is agreement that writing authentically about one's own lived experience helps construct a coherent, consistent narrative and assists in finding meaning in life's challenges and difficulties (Pennebaker 1986, 1989, 2004; Smyth 1998; Frattaroli 2008; Poon and Danoff-Burg 2011).

My own research indicates that expressive writing also offers clarity and insight, promotes emotional management through safe, effective catharsis, accelerates resolution of difficulties, and helps manage stress (Adams 2006a). In my therapy practice, I have found again and again that the journal represents an important resource for self-directed change—the process of growing oneself out of negative, painful, destructive patterns, beliefs, and circumstances. At the same time, one grows oneself into greater peace of mind, body, heart, spirit; more enjoyment and satisfaction; a quieting of the Inner Critic and the giving of voice to the wiser, gentler Inner Ally.

Although much of the formal research focuses on writing about traumatic or highly stressful matters, it is not necessary to write about deeply troubling material in order to derive the benefit. In a 2012 study, volunteers were asked to write in a structured way for a total of two hours over the course of a week about "everyday normal" problems. Nearly all made significant shifts in their attitudes about and behaviors regarding the area of inquiry (Adams 2013). Now we have the marriage of expressive writing and neuroplasticity, and for the first time we can codify methods of writing that enhance the brain's capacity to change itself in healthy and purposeful ways.

Evidence-Based Research in Expressive Writing[1]

Social science research on expressive writing[2] dates from the mid-1980s. It was then that the first landmark study was published correlating emotional release, or expressive, writing with improvement in overall health functioning (Pennebaker and Beall 1986). Over the next several years, expressive writing studies carved out a method that would hold up across three decades and hundreds of applications. Three or four segments of writing, each taking 15-20 minutes, were typically spaced one day apart.

All but a few studies included random assignments of participants to one or more experimental expressive writing groups or to a control group that wrote about deliberately neutral, mundane, or superficial topics. In most of the early studies, experimental writers were instructed to "write deeply" about a topic that was both emotionally difficult and difficult to share. Early on, the data established that writing about both content and affect (as opposed to only about the facts of the story, or only about the emotions experienced) offered the best outcomes, so experimental writers were typically instructed to include both facts and feelings (Pennebaker 1989).

Table 3.1. Adams Interpretation of Pennebaker Writing Instructions (2006a)

Write 1	Tell the story of what happened. Write without regard for grammar, punctuation, spelling, etc. Include both facts and feelings.
Write 2	Tell more of the story. Add in layers through details and parts of the story often omitted. Continue writing both content (the story) and affect (emotions).
Write 3	How has this story impacted your life? What has it meant to you? How does it continue to shape you? Are there experiences, activities or behaviors you move toward or away from as a result of this experience?
Write 4	Where do you go from here? How do you feel about this event and its impact, now that you've spent time exploring it? Is there action to take? Synthesize.

The health benefits from the initial studies were promising indeed: increased immune system function, fewer visits to health care providers for illnesses, lowered blood pressure, improved heart rate, increased antibodies for Epstein-Barr and hepatitis B patients, and other physiological improvements (Smyth 1998). Because of the cathartic nature of writing deeply about difficult, often traumatic, material, the earliest theories conceptualized the expressive writing process as a release valve that depressurized inhibition through the silent "confession" of deeply troubling material. Pennebaker and Beall wrote:

> [Statistical analyses alone do not] convey the powerful and personal nature of the majority of trauma condition essays. One woman wrote about teaching her brother to sail; on his first solo outing, he drowned. The father of a male subject separated from his mother when the subject was about 9 years old. Prior to leaving home, the father told the subject that the divorce was the subject's fault. . . . When she was 10 years old, one female subject had been asked to clean her room because her grandmother was to be visiting that night. The girl did not do so. That night, the grandmother tripped on one of the girl's toys, broke her hip, and died of complications during surgery a week later. (1986, 277).

Through the first decade of the writing experiments, it was becoming clear that the emotional expression was well-suited to the container of story, and that the arc of the three or four essays seemed to organize thoughts and feelings into a coherent, consistent narrative. Pennebaker and Seagal (1999) wrote:

[The process of constructing story] allows one to organize and remember events in a coherent fashion while integrating thoughts and feelings. In essence, this gives individuals a sense of predictability and control over their lives. Once an experience has structure and meaning, it would follow that the emotional effects of that experience are more manageable. (1243).

Poon and Danoff-Burg (2011) offered mindfulness ("paying complete attention to the experiences occurring presently, in a nonjudgmental way or an accepting stance") as a moderator that might deepen the writer's capacity to create insight and meaning. Study participants completed the Freiberg Mindfulness Inventory and other measurements before following instructions to write three times over several days, for twenty minutes each time, about a stressful experience.

Mindfulness influenced the extent of benefits produced by expressive writing. . . . [A] higher mindfulness score predicted greater change over time in decreased physical symptoms, decreased psychological symptoms, and decreased negative affect, but an increase in sleep quality and positive affect. These findings suggest that people who are more mindful benefit more from disclosing their emotions and thoughts regarding stressful experiences than do those who are less mindful (890).

Laura King, a professor of psychology at the University of Missouri and a proponent of the positive psychology movement, broke away from the trauma/illness/stress paradigm. In her earliest study, she demonstrated that those who wrote about the perceived benefits of a trauma ("the positive aspects and how the experience has benefited you as a person—how has the experience made you better able to meet the challenges of the future?") had the same physical benefits as those who wrote about loss or trauma ((King and Miner 2000, 233).

Results . . . showed that those in the trauma-only and perceived-benefit only conditions wrote in dissimilar ways. Those in the perceived-benefits-only group used significantly more positive emotion words and more words indicating cognitive insight than did those in the trauma-only group. In addition, trauma-only essays were rated as more traumatic, less positive, more negative, and less resolved than were perceived-benefits-only essays. Thus, it does not appear that the similar health results are due to similarities in writing. Essays written in the perceived-benefits condition were, quite simply, more positive than were those written in the trauma conditions. . . . [W]riting about the positive aspects of the event [may] allow participants to gain a sense of efficacy in terms of their cognitive capacities to find meaning in the event (227).

King continued her exploration of writing about self-regulatory topics with a 2001 study comparing the health benefits of writing about trauma with writing about life goals. Her findings:

Results indicate that writing about life goals is another way to enjoy the health benefits of writing without the emotional costs. Indeed, writing about one's life goals was associated with feeling less upset, more happy, and getting sick less often. . . . [T]he physical benefits . . . were equal to or better than writing about trauma, whereas writing about a traumatic life event also entailed feeling upset and experiencing lowered mood. . . . It may be possible to enjoy the benefits of writing without necessarily writing about trauma at all (805).

Burton and King (2004) asked experimental subjects to write about non-emotional topics or intensely positive experiences (IPEs) for twenty minutes each over three consecutive days. IPE writers showed increased positive mood, linguistic analyses revealed that their language reflected positive and insightful aspects, and everyone in the IPE portion of the study remained healthy. However,

none of the mediational analyses were successful in identifying the underlying mechanisms of this effect. Thus, this study, like others using the disclosive writing paradigm, indicates that writing "works"—i.e., it is associated with superior health—but does not provide a strong indication for why this is so (King 2004).[3] . . . This study broadens our appreciation for the "healing power of writing." Writing about life experience may have more general implications than have previously been explored. Furthermore, our results suggest that the discussion about the potential mechanisms of the writing effects should be broadened to include a variety of processes that are not exclusive to confronting life events (159–61).

These studies demonstrate that even a small amount of expressive writing can yield surprisingly useful outcomes; King's (et al.) in particular suggest that the outcomes need not be limited to those seeking release from long-ago traumatic experiences or current states of illness or stress. We are barely at the cusp of evidence-based research exploring writing for positive results, yet there is reason to think that more studies will demonstrate the power of writing, particularly when approached through intention, attention, and action, for positive growth and change.

Your Brain on Ink

Now it's your turn. The Inkling that follows is another structured write, a collection of sentence stems sequenced to offer extended, deepened inquiry on a common theme—an excellent way to get a fast overview of your thoughts, feelings, and perspectives on a topic. As you continue on the path to creating new neural circuitry, remember that these sentence stem templates are easy to construct for yourself. Simply write all the sentence stems at once, arrange them in a way that flows, follow with a reflection, and you've got a seven-to-ten minute writing process that will reveal the direction of your current thoughts and feelings.

Inklings

My Journal, Myself

Instructions: Complete these sentence stems with one or two sentences. Don't think too much; just write whatever comes. The first prompt in each set is for you if you already write a journal. The second prompt, in italics, is for you if you're new to journal writing. Expect to be done in five to ten minutes. Then continue to the reflection write.

Three Feeling Words

_____
_____
_____

I have been writing a journal since. . . .
From what I have heard or read about journal writing, I. . . .

My relationship with my journal could best be described as. . . .
If I wrote a journal, I would want my relationship with it to be described as. . . .

I turn to my journal when (or for). . . .
If I wrote a journal, I would probably turn to it when (or for). . . .

I don't write about . . . . .
If I wrote a journal, I would not write about. . . .

I wish I knew how to. . . .
I wish I knew how to. . . .

Reflection

Instructions: Read your sentence stems. Then give yourself a sentence or two of reflective feedback. You can start with any of these prompts:

- As I read this, I notice (or: I am aware of, I see, I'm curious about, I'm interested in, I think, I feel, etc.)
- Where do I feel this in my body? Where in my body does it land?
- Are there action steps to take? What are they?

Three Feeling Words

_____
_____

Conversation

Michael writes:

I have been writing a journal since . . . I was a young teenager, around 14 or 15; I often felt a little odd being a boy who liked to write about nature, the environment, who enjoyed writing poetry and songs. My parents never discouraged me, but the messages I got from society at large were that this activity was for girls, not boys. After all, boys do science and math and girls do the arts. I felt like I always had it backwards!

My relationship with my journal could best be described as . . . a window, a journey of freedom, a true, nonjudgmental friend who simply receives what I write and never shuts me down by talking over me.

I turn to my journal for . . . insight and answers.

I don't write about . . . being embarrassed, sex, how I sometimes use work as a drug, how being a male Caucasian fits into a rapidly changing and emerging multicultural society.

I wish I knew how to . . . develop my writing skills and personal expression like the way I have developed my songwriting skills through the repetitive use of chord progressions and various melodies.

Deborah responds: I'm intrigued by your interest in developing writing and personal expression skills as you have developed songwriting skills. Might it be that the structures and writes that support your self-directed neuroplasticity could provide the same sorts of scaffolding as chord progressions and melodic patterns offer your songwriting?

Kay responds: Thank goodness some of these gender roles from the mid-twentieth century have finally evaporated! Even late in the century, when I wrote *Mightier than the Sword: The Journal as a Path to Men's Self-Discovery* (1994), men and boys were still fighting the gender bias that writing was "just" for girls/women. I think today's boys and men have less to struggle with as the digital age and its equalizing technology have done a lot to normalize journals and expressive writing for males as well as females.

Notes

1. Adapted from Adams 2013. For a more comprehensive literature review, see the *Literature Review: Evidence-Based Research in Expressive Writing* section on page 169.

2. In the research model, *expressive writing* generally refers to the disclosure or release of emotions, traumatic memories, secrets, and other disempowering thoughts and feelings. The term is also used by psychotherapists, facilitators, coaches, educators and others to reference life-based writing as a means to growth, change, or for educators, better composition and writing skills. Therapists and facilitators sometimes use the term synonymously with *therapeutic writing* or *journal therapy*.

3. "Two strong conclusions can be made with regard to the benefits of writing. First, expressive writing has health benefits. Second, no one really knows why" (King 2002, 119).

4

The Journal Ladder

KAY ADAMS

When I started out in 1985 teaching my first journal writing workshop, I formed my theories and methods around various writing techniques. Over the years, I had adapted for my own journal many techniques from both fiction and nonfiction writing that I'd learned in creative writing and journalism classes in undergraduate school. Later, as a working journalist, I developed more techniques to help me stay on task and on time with multiple juggled deadlines and constantly colliding priorities.

A writing technique is a device that helps you match what you want to write about with a writing form that supports the intention. For instance, many are familiar with the *unsent letter*—a place to vent, grieve, confess, and fully express without the self-censorship that often results from anticipating someone on the reading end. Because the letter is, by its nature, not shared with the (non)reader, there is freedom to write in whatever way best meets the writer's need.

Another familiar journal technique is *free writing*, in which the writing starts anywhere and goes where it pleases. The idea is to "follow the pen" and see what shows up, engaging with the writing as it flows. This is, in the absence of a different idea, the default technique of nearly every journal writer. There's nothing inherently wrong with free writing; its usefulness is legendary. It is, however, only one of many choices, and it frequently isn't the most effective tool. Sometimes it can even be counterproductive, such as when a writer gets in that familiar cycle of "raging on the page," as Deborah calls it, but never moves on to actual release, relief, action, or solution.

There are dozens of techniques, each of which match up with specific therapeutic or personal growth goals. These techniques make up my basic journal therapy toolbox.

Structure, Pacing, Containment: Three Vital Tools

The Journal Ladder builds structure, pacing, and containment—three vital tools for healthy self-management—into the writing process so that techniques progress along a continuum. The lower rungs on the ladder have maximum structure, pacing, and containment; they are short, simple, well-defined, and concrete. Gradually the structure yields to embrace more fluidity, spaciousness, abstraction, and intuitive knowing in the upper rungs.

The first five techniques (from sentence stems to clustering) represent the ladder's lower rungs. These techniques invite information, comparison, and awareness. Even novice or timid writers find these structured writes accessible, and brief writing times or limited writing spaces keep content contained and manageable.

The next five techniques, the ladder's middle rungs from lists of 100 to character sketch, lean toward imaginative, descriptive, or creative elements, and they tend to generate insight. The final four techniques on the upper rungs (dialogue to free writing) have fewer boundaries and less structure; free writing, at the top of the ladder, has no structure or boundaries at all. Because the writing can go very deep without coming up for air (pacing), these techniques frequently tap into intuitive knowing or deeper understanding.

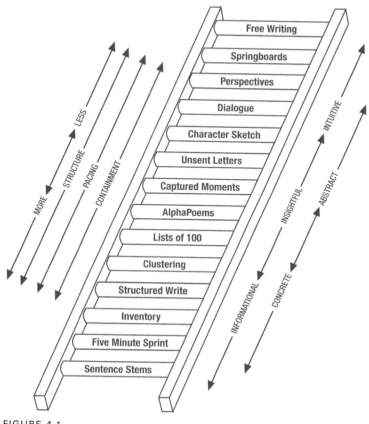

FIGURE 4.1
Journal Ladder

The Journal Ladder for Self-Directed Neuroplasticity

In this workbook, Deborah and I pair concepts of brain science with invitations for written explorations structured to reinforce the points made. We've integrated into the writing processes the very principles you'll be learning about neuroplasticity: the brain loves repetition, the brain loves novelty, the brain loves sensory engagement—all these and more are modeled in the writes themselves.

We start our writes at the bottom of the Journal Ladder. As mastery is developed, we'll move into the middle rungs, techniques that by their natures engage creativity and offer insight and connections. At the end, we'll dip into the techniques that represent the intuitive and abstract realms. When the workbook is completed, you will have experienced a preview of what might be possible when you write purposefully and intentionally over time. You will also have a template for continued writing; these processes can be repeated with new content, new stories, new experiences.

Deborah Ross

Dan Siegel (2012) describes mental health as a river of integration flowing between the banks of chaos and rigidity. When we are leaning toward one bank or the other, we lose some of our harmony and flexibility. Selecting writing strategies to promote well-being involves feeling the support of the base of the Journal Ladder, then climbing and reaching for the higher level techniques.

This is not necessarily a vertical progression up the Journal Ladder. Rather, it is a constantly recurring choice of the best writing technique or prompt for where you are on any given day, and where you want to place your intention, attention, and action.

Siegel further suggests that mental health is an integrated flow characterized by the acronym FACES: flexible, adaptive, coherent, energized and stable. Another way to use the Journal Ladder is to select writing techniques and prompts that are in service of that flow, moving up and down the rungs according to the techniques and prompts that might best support flexibility, adaptability, coherence, energy, and stability—again based on where you wish to place intention, attention, and action.

Kay's Glossary and Guide to Techniques on the Journal Ladder

*Asterisked techniques are practiced in this workbook

Lower Rungs

Sentence Stem.* A sentence-completion process. Fill in the blank with a word or phrase. May be openly universal (*Right now I feel ———*) or highly customized to an individual's immediate question, problem or interest.

Five-Minute Sprint.* A timed writing process designed to bring focus and intensity in short bursts. Excellent for those who are new to journal writing, who are uncertain about how to start, or who state they do not have time to write journals.

Inventory. An assessment of life balance in major areas of living (health, family, home, work, spiritual/religious, emotional well-being, etc.) Gives a quick picture of which life areas might need attention.

Structured Write.* A series of sentence stems grouped and sequenced to reveal consistently deepening layers of information and awareness.

Clustering.* Visual free-association from a central word or phrase. Lines and circles connect key thoughts and associations to the central core. Work quickly to maximize results. A brief writing to synthesize findings may follow (Rico 2000).

Middle Rungs

Lists of 100. A list of 100 items, many of which will probably be repetitions, on a predetermined theme or topic. Repetition is an important part of the process. Topics can be about any current issue (for example: 100 Things I'm Sad About; 100 Things I Need or Want to Do; 100 Places I Would Like to See). At the end of the list, group the responses into themes and synthesize the information.

Alphapoem.* Write the alphabet, A–Z, or any collection of letters, vertically down the side of a page. Then write a poem in which each successive line begins with the next letter. Excellent for groups as it promotes a high level of participation and sharing. Adolescents and reluctant writers respond well.

Captured Moments.* Vignettes capturing the sensations of a particularly meaningful or emotional experience. Written from the senses with strong descriptors. Captured Moments of beauty, joy, blessing, or calm can add balance, hope and perspective to a challenging time.

Unsent Letters.* A metaphoric communication to another that is written with the specific intention that it will not be shared.

Character Sketch.* A written portrait of another person or of an aspect of the self. Can also be written about emotions by personifying an emotion and giving it a characterization—an appearance, a style of dress, a personality and temperament.

Upper Rungs

Dialogue.* A metaphoric conversation written in two voices. Anyone or anything is an appropriate dialogue partner. There is no constriction by time, space, physical reality or literal voice.

Perspectives.* An alteration in point of view that provides a different perspective on an event or situation. May be alteration in voice (write in third person instead of first person) or time (write in the future or in the past).

Springboard. A free-write with a prompt. Starting a free-write with the smallest structure of a question, thought or topic can focus and frame the writing session.

Free writing. Unboundaried, unstructured, unpaced narrative writing. Useful for creative flow or spontaneous writing sessions. Can be structured by adding a time limit or page limit.

5

Brain Maps

> We are what we repeatedly do. Excellence, then, is not an act, but a habit.
>
> —Aristotle, Greek philosopher and scientist, c. 340 B.C.
>
> My experience is what I agree to attend to. Only those items I notice shape my mind.
>
> —William James, an early thinker in modern psychology, 1890

DEBORAH ROSS

Neuroplasticity is such a new word that my computer flags it with a red dotted line every time I type it. The word might be new, but the process isn't; the brain has probably always been neuroplastic. However, science is now able to validate what poets, philosophers, and positive thinkers have intuited as far back as Aristotle: *We can change our brains by changing our minds. We can change our minds through the power of our intention, attention, and self-directed action.*

The way the brain does this is by creating new neural maps, or circuitry, that redirect "traffic" toward the dominant thought, emotion, belief, or experience.

London cab drivers must memorize the unusually complex street map of the city before they can pass a rigorous licensing test. A study (Maguire et al.

2000) was done in which the brains of aspiring cabbies were scanned before training and after taking the licensing exam. The brains of a control group who were not in taxi driver training were also scanned.

The hippocampus is the area of the brain that creates visual-spatial memories. It can be cultivated and strengthened with new neural maps. The hippocampi of cabbies who passed the test and were licensed were shown to be more well-developed, or thicker, than either the control group or those drivers who had failed to memorize the London street system sufficiently to pass the test.

In 1992 the Dalai Lama offered neuroscientist Richard Davidson, founder and director of the Center for Investigating Healthy Minds at the University of Wisconsin-Madison, the opportunity to scan brains of Tibetan monks to learn if years of cultivating well-being created changes in the brain. The monks had lifetime practices of mindfulness meditation with focus on compassion and loving kindness. Their functional MRIs showed very well-developed *insula*, the region of the brain that is activated by focus and attention on the body and on feelings. This finding also extends to the brains of those who have training in Mindfulness Based Stress Reduction, a program designed by Jon Kabat-Zinn (Davidson and Begley 2012).

The technology now exists to see what has been happening all along. Focused attention over time manifests changes in the brain; the brain makes new neural maps. Like a body builder whose muscles are still strong even as they are at rest, so too does the brain retain its strengthened areas, even while these areas are at rest. However, like a body builder who stops exercising and grows flabby, so too does the brain lose strength in developed areas if those circuits are no longer being asked to fire regularly.

What sages and scientists, athletes and artists, have had inklings of all along is now visibly measurable. We have choice over the level of excellence in our lives; how we focus our intention, attention, and action creates new maps in our brains.

Kay Adams

Writing makes a map, and there is something about a journey that begs to have its passage marked.

—Christina Baldwin (2007)

At a *Your Brain on Ink* workshop, a participant showed Deborah and me her journal. It was a joyful collage of writing, art, newspaper headlines, ferry tickets, thank-you notes, charts, graphs, cartoons—a visual journal that celebrated every inch of her life. It was a glorious integration of self, a road map of her relationships, passions, interests, and the world in which she fully lived.

My friend Brenda is a book artist and writing teacher. Brenda coaxes unique expression out of all of her students, by helping them make beautiful journals—sometimes creamy pages stitched into supple leather covers, sometimes a simple folded-paper book with a pocket. Writing in a book made by hand, her students say, is an intensely pleasurable experience, an invitation to what Brenda calls the "voiced life."

For most of the 1980s I wrote in wide-lined spiral notebooks, and I learned to save the last eight pages for charting the themes, connecting the dots, finding the patterns, noticing the rhythms.

Writing makes a map. Is it such a stretch to think that the map made by writing is charted in the brain?

The Inkling that follows invites you to make a map of your own writing journey. As we'll learn, the brain loves novelty, and your journal loves maps, lists, doodles, and other forms. Play with this map-making process. Then stand back, take it in, and write a reflection.

Inklings

Making a Map

Instructions: Create a map of your journey with writing throughout your life. Make this map in any way you wish:

Three Feeling Words

_____

_____

_____

- A visual map using collage, illustration, colored markers, or other artistic expression
- A bullet list of turning points and significant life events
- A narrative story of your journey with writing
- Or any way you want.

Begin at any age or development stage you choose. As a reminder, there is no "right" way to do this, only the "write" way.

Reflection

Step back from your map and reflect. What do you notice? What do you see? Where do you feel it in your body? Are there any surprises?

Three Feeling Words

_____

_____

_____

6

The Reflection Write

KAY ADAMS

By now you have practiced reflecting on your writing after it is complete. Perhaps you've harvested insight or even an "aha" moment.

The reflection write is the key to developing the observational muscle necessary to notice what is happening as a function of writing. The reflection write harvests insight, codifies thinking, and brings detached mindfulness to the writing process. It comes at the end of each writing session in the form of one or two sentences about the writing just completed, in response to prompts such as, *As I read this, I am aware of. . . .* or *What did I notice in my body as I was writing?* or *What is an action step I could take to move my intention and attention in the direction of what I want?* (Adams 2013).

When you write, your mind can be scanning/projecting into the future or reviewing/ruminating about the past. Your body, however, is in real time. Tracking the embodied experience of writing helps you stay grounded in the present moment.

There is growing evidence that resolution of trauma (or its smaller cousin, stress) requires somatic (body) involvement. The reflection write develops the habit of checking in with the embodied experience of writing, a good gauge of how your nervous system is processing the writing.

When you are present to what emerges on the page by reading what you have just written and writing a few sentences about what you notice, you are developing an observational part of your brain. You are gaining separation from the write itself and taking note of both the process of writing as well as any insights that the writing yielded. *What happened in my body as I wrote? Did my handwriting change? Was there a smile on my face or tears in my eyes? Were there any "aha" moments?* Paying attention, cultivating curiosity and noticing what emerges as a function of the process of writing is similar to meditation practices that cultivate concentration and invite insight.

There is a page of reflection write starters at the end of this chapter.

Deborah Ross

A growing body of research provides evidence that our brains change based on how we direct and focus our thoughts. The reflection write is the consummate expressive writing tool for focused attention. A typical journal entry may be written quickly, often in five to seven minutes or even less time. As the write is concluded, it is read so that the experience can be taken in. We are noticing from an observational platform; therefore we are flexing observational muscles. Attention is directed both to the content of the writing and the writing process.

The reflection write is an exercise in the process of paying attention, particularly if curiosity and compassion are brought to that process. Reflection supports the process of choosing to fire circuits that over time will change our brains in service of greater healing—and, likely, the authorship of a more integrated, coherent story.

Reflection Write Starters

At the end of each writing session, read what you've written and give yourself a sentence or two of feedback. Choose from any of these approaches, depending on what best fits the moment.

- As I read this, I am
  - aware of. . . .
  - curious about. . . .
  - noticing. . . .
  - surprised by. . . .
- The integrative somatic experience
  - What happened in my body as I wrote? Where did this write "land" in my body?
  - Did my handwriting or keyboarding change?
  - What did I notice emotionally?
  - Any "aha" moments? Where did I feel them?
- Action orientation
  - Is there action to take? If so, what?
  - What is my next step?
  - How does this learning inform my current reality?
  - Where can I best place my intention? My attention? My action?
  - What is one thing I can do today?

II
THE WRITE WAY TO POSITIVE BRAIN CHANGE

7

The Brain as Velcro and Teflon

DEBORAH ROSS

Here's the bad news: self-help is an uphill climb. We are wired to embrace negativity.

Our brains are "Velcro for the negative and Teflon for the positive" (Hanson 2013). Why does it make sense to have our brain, like a powerful vacuum cleaner in our skull, scanning for, sweeping in, and fiercely holding onto the negative?

Neuroscientist Jaak Panksepp (2012) suggests that our brain represents a kind of living museum. It contains all of our cavemen/women ancestors, beginning with Neanderthals, through Cro-Magnons, along with ancient versions of all of the creatures from the sea to the land—jellyfish to lizards to gorillas. In those times where survival, not quality of life, was the order of the day, the operative question was not, "What am I going to have for lunch?" but rather, "Am I going to be lunch?"

Our evolutionary history, encoded in our brains, primes us to learn as quickly as possible from bad experiences. Those learnings ride an express track into the brain's storage centers; we are set up to learn much faster from pain than pleasure:

- Couples therapist John Gottman (1994, 1999) reports that in healthy marriages a ratio of five positive comments to one negative keeps the marital eco-system robust (1994). He also suggests dedicated blocks of time—no multi-tasking!—that are devoted to admiration, appreciation, and affection. It is vital that we take in the good (1999). It is common for couples to struggle with how to offer a fully positive appreciation for their partner or take in a fully positive expression of love and caring.

- We tend to live in a zone where we are more inclined to language something in the negative—all that we do not want or are not receiving. Even positive expressions are often weak and merely reflect the absence of the negative. Progress in therapy, for instance, is often expressed as, "I'm not as stressed/anxious/depressed today."
- Should we be offered a lovely gift of gratitude or praise, we often hiccup or deflect rather than embrace the gift by fully taking it in. "Yes, but," is the phrase that often drives our relationships with others as well as the one we have with ourselves. We respond like non-stick Teflon for the good that comes to us.

Our evolutionary, ancestral wiring has rendered us highly skilled at retaining the learning lessons from bad experiences and has left us rank amateurs at retaining the learning from positive experiences. In addition, our radar for the negative also contributes to how we store not only the memories of experiences but the translation of these experiences into expectations, assumptions, overall beliefs about ourselves and others, including a sense of who is "other" in our world.

In the end we have an imbalance in favor of the negative, which was useful for survival under harsh and threatening conditions. That imbalance includes both an internal radar sweep for impending negative experiences, an amygdala that sets off its own alarm bells with attention-getting urgency, and a rapid encoding of those experiences in our brain. We go from negative mental state to negative neural trait with dazzling efficiency.

Kay Adams

We'll follow up our sentence stems and structured writes with the five-minute sprint. (If your personal preference is to have a slower pace, feel free to rename this the ten-minute trot or the seven-minute stroll.)

Seated at the second rung of the Journal Ladder, the five-minute sprint is just what it sounds like: five minutes of fast writing. Keep the pen or fingers moving. Don't stop to edit or read. Just keep writing.

You'll likely be surprised at how much coherent writing emerges in a spare five minutes. Precede it with a minute of closed-eye deep breathing and follow with a read and reflection write, and you've got an oasis of clarity or relief in about eight minutes—a time unit most people agree is manageable even (especially!) on the busiest days.

Be sure to have a timer available, and use it. With the reflection at the end, you'll spend about twelve to fifteen minutes on this process.

Inklings

Velcro and Teflon

Instructions: Set your timer and write for five minutes on each of these two topics. Then reflect on the next page.

Three Feeling Words

_____

_____

_____

My brain grabs like Velcro for. . . .

My brain is non-stick like Teflon for. . . .

Reflection

What do you notice about your brain's tendencies to grab (Velcro) or let slide (Teflon)? Reflect on your findings.

Three Feeling Words

_____

_____

_____

Conversations

Damaris: I noted that my Velcro brain, while brilliant at snatching at everything shameful and humiliating and saddening, is also quite good at snatching at valuable information, information that has allowed me to have a career.

Deborah responds: Thank you for this reminder. Velcro in and of itself is neutral; it's what it grabs that can be problematic. Velcro also grabs useful stuff, which demonstrates that we can actually make the positive stickier.

Ann: My brain grabs like Velcro for not only negative, unpleasant experiences—large and small, but also grabs for experiences that are basically neutral ones and then comes up with (or down with?) any negative situations that might possibly develop from the neutral ones that actually happened. So I tend not only to perpetuate negativity— I create it.

Deborah responds: We are so creative with our ability to turn neutral to negative, aren't we? If we don't have enough on our plate, then let's get pre-emptive and proactive about creating more!

Jack: There seemed to be a direct correlation between my negative Velcro things and the Teflon things that *could* but *don't* override them (e.g. Velcro for "who do you think you are?" and Teflon for "you've got some real possibilities, dude!") Then I get into the whole debate about who's right and who's wrong—usually ending up on the self-judgment ("don't be so full of yourself") and then I crater.

Deborah responds: Oh, self-judgment—ouch! Another Velcro attachment. It's impossible not to be influenced by the negative; we are hard-wired to embrace it. Remember, this is our survival that we're talking about!

Please remember this is just a snapshot in time. You have a whole internal movie's worth of memories and associations. While the negativity bias in the form of Velcro is there, you may notice if you do the writing process again, some positives may emerge that didn't come up the first time. Being present to the positive, noticing what is going well, and paying attention to your Teflon moments through writing—that all adds up to a wonderful start.

Shannon: In doing the writing, I almost felt different parts of my brain activated. Is that possible? So much flowed from the Velcro side, and with such emotion! The Teflon side felt so purposeful and like sorting through items in a closet.

Deborah responds: The cognitive dissonance of "neurons that fire together, wire together" versus the primal bias for survival loads a lot onto a circuit. Whether or not it's a physiological event, it's certainly possible that you could be translating a "felt sense" as you activate different circuitry as different parts of your brain light up.

Vivien: I'm not proud to say that the strongest Velcro is around the things I say to myself about what I could have done better, what I should have done with more dedication, etc. I Teflon (if I may use that as a verb) the moments that I could really support myself and give myself credit for a job well done, whether that's some good writing, a strong relationship I've cultivated, or any other accomplishment. A quick example that comes to mind: someone recently said to me, "Wow, you ran a marathon?" and I answered, "Yeah, but that was a long time ago, and I've only run one."

Deborah responds: The research has suggested that "Tefloning" recognition of our accomplishments or compliments that come our way is something that women are more likely to do. We seem to have greater deflector shields. Neurologically, it has something to do with our desire to remain in affiliation, to be part of the group. From ancient hunter/gatherer days when women were more likely to be gathering than hunting, women would be in a group—safety in numbers also ensured the children would be tracked and tended to. So women often struggle more with owning and taking in accomplishments as we don't want to be different (separate) from the group. As well, there are plenty of societal messages about that too regarding modesty and humility. Congratulations on running a marathon!

8

Your Limbic System

DEBORAH ROSS

How do you create a more Velcro-like experience for the positive? How do you use all of this information about the ways your brain is primed to ensure that you survive, but is rather disinterested in whether you thrive? What is helpful? How does expressive writing help construct a lasting story with headlines that are encouraging, healing, and reflect a vision of the kind of life you wish to lead?

The short answer is: Make positive things stickier.

In the service of that mission, we turn to two sections of the brain, both part of the limbic system. One of them, the hippocampus, has a significant role in appraising and consolidating information from short-term to long-term memory. It turns out that the hippocampus loves novelty; it's much more likely to hang on to something that has surprised, engaged, or awakened it. The hippocampus is also quite vulnerable to the stress hormone cortisol. Neurons can be weakened or killed in the presence of cortisol, particularly if it is part of the daily soup served in the brain. That toxic meal will actually shrink the hippocampus itself.

However, the hippocampus is also a primary site where neurons are born, and their survival odds are enhanced by the presence of novel experiences. When we are playful, spontaneous, and inviting new experiences or new ways of noticing our experience, we encourage the brain to grow (Hanson 2013).

The well-being of the hippocampus makes it more likely that the alarm bell sounded by the adjacent amygdala will be appropriately evaluated and perhaps modulated. The amygdala's job as fire alarm for our system is to get us to "act before we think," and the hippocampus, when it is healthy, is able to use memory to question whether this level of urgency is really appropriate for the situation. A key role for the hippocampus is, therefore, its ability to assess the context in which the alarm is screeching.

When our neurons are firing, and all circuits involved in the activity are a "go," then the genes in those cells are primed to become more active. That activity can invite two processes, one that yields better synaptic connections between neurons, making communication more efficient; and one that can increase the insulation around the axon, or the tail of the neuron. Increased insulation makes it possible for the neuron to work so efficiently that its functioning can improve as much as 3,000% because impulses travel much faster. In addition, the recovery time of the nerve is shortened so it can fire again sooner (Siegel 2012).

These neural findings eminently support the importance of practice, practice, practice. Those who have been recognized for their achievements in music, art, athletics, undoubtedly have beautifully insulated nerve axons in the regions of the brain that mediate their unique skills.

Practice can also refer to having a practice that is in service of the skill of well-being. Richard Davidson has spent years studying how our brains change in response to how we focus our attention. Beginning in 1992, when he started studying the brains of the Dalai Lama and his Tibetan monks, and continuing to the present time, Davidson has been working to understand how mindfulness-based practices can contribute to our well-being, noting how our brains change in the process cultivating these practices (Davidson 2012).

Having a regular journal writing practice, which is a concrete and verifiable form of attention, shares common features with mindfulness-based meditation practices. Even when the writing is free-form, with no particular objective, the choice of writing as an activity is intentional. In picking up the journal and pen, or opening the screen of a digital journal, a choice was made to engage the brain in the process of life-based writing.

Kay Adams

This workbook is abundant with many writing techniques and prompts that invite you to direct and focus your attention in intentional and purposeful ways. Responding to any journal prompt invites attention and focus on that theme. The reflection write that follows brings the writing in line with mindfulness-based meditation practices; it offers compassionate observation and an opportunity to harvest insight and apply it in real time.

Purposeful and intentional writing can manifest observable results in a surprisingly short time. One obvious reason is the paper trail that writing produces; it is possible to literally read your own mind when you write it down. Another reason is because the act of writing—whether it is by hand or on a keyboard—moves thought, feeling, and emotion through the mind/body and onto the page or screen. This embodiment of an interior state in the exterior world—words on paper or pixels on screens—is itself an ongoing dress rehearsal; we are living our way into embodiment of healing, growth, and change, as evidenced by the self-reporting that a journal provides. We are anthropologists on an archaeological dig of our own lives.

Just for Fun

The Amygdaloids: A Heavy-Mental Band

The Amygdaloids are scientists who shed their scientific garb at night and take to the stage with songs about love and life peppered with insights drawn from research about mind and brain and mental disorders.

Their first CD, released in 2007, was called *Heavy Mental*. *Theory of My Mind*, their second CD, was released in 2010 and features Grammy Award winner Rosanne Cash on backing vocals for two songs ("Mind Over Matter" and "Crime of Passion"). The band's third set of recordings was an EP called *All in Our Minds*, in which each song had "mind" in the title.

NYU neuroscientist Joseph LeDoux, a pioneer in the study of the brain mechanisms of emotion and memory, is the lead singer, songwriter and rhythm guitarist. He is the best-selling author of two books, *The Emotional Brain* and *Synaptic Self*. The lead guitarist is Tyler Volk, professor of biology and science director of the environmental sciences program at NYU and author of *CO2 Rising, Metapatterns, Gaia's Body*, and *What Is Death?* Daniela Schiller was a postdoctoral fellow in cognitive neuroscience at NYU when she joined the band as the drummer, but is now an assistant professor of psychiatry at Mount Sinai Medical Center. On bass and vocals is Amanda Thorpe, who studied neuropsychology at University College London and currently works in advertising. Amanda has an independent musical career, her most recent project being the CD *Promenade*.

Source: www.amygdaloids.com

9

The Brain Takes the Shape the Mind Rests Upon

DEBORAH ROSS

One of the core tenets of Rick Hanson's (2011) work is the brain takes the shape the mind rests upon. By this he means that the physiology of the brain, as seen on functional MRIs, is different for those whose thoughts are associated with compassion and loving kindness (e.g., Tibetan monks) than those whose thoughts are associated with fear and pain (e.g., people with current or past trauma).

You might notice that this idea—the brain takes the shape the mind rests upon—makes a distinction between the brain (physical organ) and the mind (focus and attention). Certain regions of the brain, in all its powerful circuitry, are shaped by the mind. Our predominant thoughts inform the shape our brain is taking at any given time.

This is such a novel concept that we may not be fully able to wrap our minds around it without taking pen or pencil to paper. Kay will tell you more.

KAY ADAMS

This chapter contains two writing processes, a sentence stem warm-up and a new technique, clustering.

This is the third time you've experienced the technique of sentence stems, and by now the repetition has probably offered a certain familiarity and perhaps even mastery: hey, I can do this! As Deborah has already mentioned, the brain loves repetition; it helps make the positive stickier. The journal loves repetition, too; your favorite techniques, used over time, are like climbing into your favorite sweatpants—familiar, cozy, comfortable.

One thing I love about sentence stems: they give us a baseline against which we can measure movement, process, and understanding. If you were to complete the two sentence stems every month for a year, you'd likely see evidence of deepened embodiment of the brain-changing practices you will work with in this book—assuming you continued to work with them.

This brings us to a painfully obvious truth. Much of the investment in your new neural pathways will not stick around unless you are reasonably vigilant about keeping up the practice. In this sense, *practice* is both a noun and a verb—both the *being* and the *doing*. As my yoga teacher once told me when I griped about not feeling flexible, "Kay, there's a reason they call yoga a practice. You have to practice." Like yoga, your best benefits will come from incorporating purposeful and intentional writing into your lifestyle. It's not necessary to write every day. It is necessary to write.

Ideally you'll be able to commit at least an hour a week. That's manageable, right? One hour a week? You can keep your circuits buzzing with three or four weekly writes of fifteen to twenty minutes each. Benefits also accrue from shorter, more frequent writes, such as constructing a sentence stem or two about your situation (e.g. *The most important thing to do—*; *What I want to say—*; *My next step—*; *The truth is—*) and spending five minutes responding.

You get to decide whether you prefer longer (but still time-efficient) writes three or four times a week, or more staccato writes any time you can grab five minutes. Either way, when combined with the self-directed neuroplasticity techniques in this book, you'll be laying down circuitry as you go.

Inklings

What's on My Mind?

Instructions: Complete the two sentence stems, then continue on to the next Inkling.

Three Feeling Words

_____

_____

_____

"The brain takes the shape the mind rests upon." Right now, to me, that means. . . .

Some things on my mind right now are. . . .

Charlotte: Some things on my mind right now are. . . . dendrites, lobes, neurotransmitters . . . I've been attending a Brain-Based Instruction workshop all day. Also, memory consolidation and learning; vulnerability to illness, injury, negative thinking; critical thinking, curiosity, imagination.

Julia: Some things on my mind right now are . . . my mental to-do list, stress, stuff I have no control over, negative thoughts, anxiety; also nourishing things, peaceful things, renewal.

Kay Adams

Let's continue our exploration of "the brain takes the shape the mind rests upon" with the clustering technique (Rico 2000). Often called mind-mapping or webbing, this is a graphic journal technique that lets us see a visual representation of thoughts, associations, and connections. Clustering is part of the lower rungs of the Journal Ladder; thus, it is reliably structured, stable, contained, and yields information. Because it is at the transition point between the lower and middle sections of the ladder, it also hints at the creativity and insight that the middle rungs are known for. The visual form is key to the technique's creative/insightful capacity; there is often sudden or startling awareness in how the diagram looks on the page, which words and thoughts are juxtaposed, the connections between seemingly disparate things.

When you've finished your cluster diagram—because you've run out of space, or you've run out of juice, or the cartoon lightbulb over your head just switched on—you can step back, observe, and write whatever comes as a reflection. Sometimes this takes the form of jotted notes about the various arms of the cluster diagram, or it may be an "aha" moment that needs capturing before it wisps away. Synthesize your visual clues and cues in the reflection write.

Inklings

My Brain

Instructions: Following the approximate pattern of the starter cluster below, make a "My Brain" cluster on a full sheet of paper. Starting from the center circle, map out everything your brain is habitually paying attention to. Family, work, home, concerns, successes, activities—what is dominant in your life as you are living it today? Free-associate using lines and circles, always coming back to the center when you begin a new thought. When you have completed your cluster, stand back and observe, then continue with a five-minute reflection write to synthesize your findings.

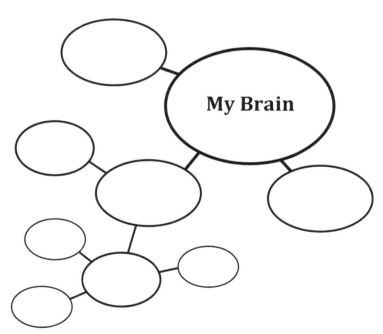

FIGURE 9.1
"My Brain" Cluster

Reflection

Instructions: Review your cluster and notice what is obvious, curious, perplexing, or otherwise interesting. When you are ready to synthesize your learning, set a timer (phone, watch, kitchen stove) for five minutes. By hand or keyboard, write as fast as you can about what you have noticed. When the timer signals, finish your thought and stop writing.

Consider these questions. Choose one that engages you, or just start writing.

- What is the dominant energy or information in the cluster?
- What is the shape of your cluster? Balanced? Jagged? One whole section left empty?
- What does your cluster tell about the shape your brain is taking?

Three Feeling Words

_____
_____
_____

Conversations

Vivien: I found that my clustering was full of opposites, and in that I saw the gift of a refusal to be contained. My writing afterwards also took a shape I did not expect—literally, I wrote all over the page. In the whole exercise, I found freedom in the reminder of how valuable it is to treat my brain (and myself) with kindness and generosity.

Deborah responds: Since the phrase is, "the brain takes the shape that the mind rests upon," it sounds like your cluster reflected that refusal to be contained. That's a nice gift to yourself as you notice all the opposites or polarizations.

Damaris: I have an uneasy relationship with my brain. It seems that I have been treating my brain as if it were an Oldest Child and must be responsible for all the younger sibling parts of me that need tending while I free-float through my life. I see that my brain, which is on board with me regarding our goals and plans, needs some help readjusting, retraining some of my less helpful attitudes and retraining life-long habits that kept me out of trouble by keeping me on a tight leash.

Deborah responds: It's common for people to think of their brain as just the seat of all the executive functioning, management and responsibility wielding. If you have a movement meditation practice, such as yoga, tai chi, or stretching, you might want to create another cluster after engaging in that activity. Do your reflection, then compare the two reflections and note any differences. And you can engage your brain by gently relaxing that tight leash a little, and intentionally practicing the "retraining." These are the thoughts, actions, and behaviors that will help create newer circuitry.

Damaris continues: I did the first cluster in the late afternoon after a busy day. The second cluster I did soon after rising, following thirty minutes of stretching. The differences amazed me. The first cluster is, visually, an awkwardly formed crystal, as if crystal shards had been broken off by a vandal. The second clustering is, visually, a blossoming flower, with petals symmetrically forming around the ovule. The first one looked for logical connections (what does that crystal assert about my capacity to reason?); the second one moves along with intuitive and metaphoric associations.

Deborah responds: How exciting to notice such a profound difference! Further evidence that in our aliveness, our brain manifests multiple pictures that reflect its current state. And a cluster tomorrow will undoubtedly have some recognizable webbing but will also have its own revelation.

10
Neural Darwinism

DEBORAH ROSS

Just as our physical fitness follows the "use it or lose it," model, so does our brain. Neurons that have fired together repeatedly are less likely to be lost to disuse and the brain's ability to prune "dead wood." Conversely, areas that are repeatedly and consistently used can thicken and grow strong, as was demonstrated by the scans of the brains of London taxi drivers and Tibetan monks that we learned about earlier.

This phenomenon is known as neural Darwinism, or "survival of the busiest (brain circuits)" (Hanson 2013).

Until adulthood is reached, the brain periodically takes a look around, assesses the most- and least-used circuitry, and prunes circuits that aren't used much. As a child grows and matures, the brain is wiring circuits that are mediated by the relationships with parents, siblings, grandparents, teachers, and others. When those relationships are characterized by love, attunement, presence, and a general sense of well-being, the child's experience of the world is likely to be that it is safe and trustworthy, and the child will likely have circuitry for empathy and compassion. The negativity bias is still there because survival comes before well-being, but there are also positive connections with others and the capacity for that connection.

However, let's take an already installed negativity bias, add negatively reinforcing circumstances in childhood and adolescence—emotionally unavailable parents, neglect, or abuse are just three possibilities—and mix them together in a brain that is pruning and shaping those neural trees. By the time full adulthood has been reached in (for neurological purposes) the mid-twenties, the brain will be wired to fire on that bias. Since the brain likes the familiar, although it learns better with novelty, the owner of that brain will likely continue down the negatively biased path of least resistance. In its extreme version you may have an adult whose circuitry for positive connection, compassion and empathy is quite sparse.

The good news, of course, is that a new neural train can be routed down a different track. Those circuits that fire regularly and are the busiest in the system will be the stronger ones. As you can see, it pays to cultivate the positive pathways so that they are the ones that are well-used, best maintained and most efficient at firing.

Bottom line: use them; don't lose them!

Kay Adams

For decades I have been telling my therapy clients that, despite what the bumper sticker says, it *is* too late to have a happy childhood—but it is *not* too late to have a happy, healthy, and productive adulthood. Now I know why: neural Darwinism!

In the 1990s I had a full-time job as the journal therapist for a psychiatric hospital that specialized in the treatment of a complex mental health disorder; the patients' average length of in-patient stay was about seven weeks. It seems unthinkable in today's managed care environment, but that's how it was, back in the day.

The length of stay, plus the three mandatory journal therapy groups and one private session that all patients attended each week, gave me an opportunity to witness how an expressive writing habit develops over time. I focused particular attention on those who were new to the writing process; I took care to help them structure a regular writing routine, beginning with the three weekly groups and extending to follow-up writes and reflections after individual and group therapy sessions, suggestions for processing difficult emotions, and tips for celebrating victories in their journals. Most patients embraced this structure; others attended groups but didn't write much in between.

My discharge interviews included an examination of how well the patient had integrated writing, not only as a part of the treatment process but also as a part of a larger personal process. Had they developed a journal habit that would support them as they returned home?

In the late 1980s and early 1990s, the conventional wisdom was that it took about forty-two days (six weeks) of repeated, consistent practice to habituate behavioral change. Prior to this hospital job, my anecdotal evidence suggested that a new journal writer could establish an ongoing writing habit if s/he wrote about three times a week, for about twenty minutes at a time, with pause for reading and reflection, over at least six weeks' time.

Therefore, I hypothesized, most of the patients who had attended facilitated writing groups three times a week for at least six weeks would likely have developed a personal process habit.

I was wrong.

It turned out that those patients who had *only* attended writing groups and an individual session with me each week, but had not written privately, reported that the writing had helped them, sometimes dramatically, in their treatment process. Although nearly everyone said that writing had been a significant factor in their hospital stay, the majority did not know if they would continue writing after returning home. In the absence of a facilitator and a writing community, they weren't sure if it would have the same impact.

Patients new to journaling who attended three writing groups and one individual session each week *and* wrote regularly in their journals on their own time had much different outcomes. They reported being strongly bonded to their journals, considered the journal a primary tool in the recovery process, and when asked to choose words or phrases to describe their relationships with their journal at discharge, regularly chose words such as *comforting, helpful,* and *friend in need.* These were the same or similar words as those chosen by patients who were experienced journal writers when they began treatment with us, as expressed in their journal therapy intake assessments.

I have long been a proponent of *not* writing every day unless it is in the service of a specific practice, such as a prayer journal, a medication/side effect log, or the first thirty (sixty, ninety) days of abstinence from drugs, alcohol, or other substances. The reason I discourage the idea of writing every day is two-fold: first, it sets up an unrealistic expectation; a "miss" of a day or two is considered a failure. Second, I don't think daily writing is necessarily more effective. A minimum of three times a week, accompanied by regular reflection to harvest insight and anchor fledgling beliefs and perceptions, seems to be sufficient to create and sustain a journal habit.

I still think twenty minutes three times a week is great if you can do it, but I no longer consider it the gold standard. If you can commit as few as five to fifteen minutes per write, several times a week, it will likely get you attuned. Longer writes, of course, allow you to explore more aspects of any given story, but three sequential ten-minute writes are not necessarily less effective than one thirty-minute session. It is, however, important that the read and reflection be completed for each session; that's where a lot of the integration and installation takes place.

In the service of your own neural Darwinism, Deborah and I suggest that you write in the same journal at least three times a week for the next six weeks. Work your way through this workbook, and then continue writing using the techniques, tips, and prompts you've learned. Keep those circuits busy forming a new pattern of positivity, resilience, and optimism. What have you got to lose except some old, worn-out tracks?

11

Neurons That Fire Together Wire Together

Deborah Ross

Neuroscientist Carla Shatz (1992) coined the phrase "cells (neurons) that fire together wire together." Her statement reflects the work of Donald Hebb who, in 1949, wrote about synaptic plasticity (Hebb 1949).

Neurons do not literally fire at exactly the same time, resulting in wiring. Rather, neuron A might activate B which activates C. At the same time that some neurons are activated, others are deactivated or turned off.

Essentially, the brain builds synaptic connections between neurons in response to a triggering event. The electrical current takes a particular path through the neurons. And the connections between neurons will be strengthened in the presence of novelty, repetition, an emotional experience, or when our attention is focused.

This happens so quickly that it might seem like a simultaneous firing, but the circuit board of our brain is not lighting up all at once. There is travel time from one neuron to another. But the stronger the association, the faster the firing. Remember that for those who have committed to the discipline of focused intention, attention, and action—elite athletes, accomplished musicians, scholars—the efficiency of firing can be as much as 3,000% faster as a result of the repetitive firing of the circuit! (Siegel 2012)

We have discussed how the brain has a negativity bias; we're just wired that way. However, the process of making neural maps—new neural circuitry—is the same, whether the path laid down is for resilience, harmony, and pleasure, or for anxiety, agitation, and despair. Here's an example:

Consider a pile of sand at the beach. Pour a bucket of water on it, and a path is created through the sand pile. Pour another bucket of water, and the path is deepened.

Now consider that a third bucket of water is a wash of emotion, a primary way that neurons start firing. Emotional context is part of how the brain wires, particularly when associated with relationship. Perhaps you can begin to glean how our emotions, and particularly the emotional washes we get when we feel in relationship with someone or something (nature, a creative project, a passionate purpose), can be key factors of how the brain lays down circuitry.

Other components that help neurons fire together are the addition of sensory experience, the capacity to sustain attention and focus, and integration of mind, body, feeling, and spirit—all with frequent repetition. We'll be covering all of these ideas, and more, in upcoming chapters.

Kay Adams

It's exciting, isn't it? That we have the power to change our brains through our conscious thought and action? And that writing can be one of the portals through which we gain entry?

It feels somewhat magical, and that's a perfect reason to use a fantasy writing prompt that engages the novelty of *let's pretend* with a sensory wash of feeling.

We'll combine the floaty nature of the write itself with a form that holds the string on the balloon: the familiar five-minute sprint.

On the next page, you'll be guided in an imagery process. Bring all your senses to the experience! You'll write for five minutes, then read and reflect. Conversations follow.

Inklings

A Wash of Feeling

Instructions: Read the guided imagery to yourself. Then close your eyes, take some deep breaths, and write for five minutes by a timer. When the timer signals, finish your thought and wrap up. Then read what you've written and proceed to the reflection write on the next page.

Three Feeling Words

_____

_____

_____

Imagine that you are at the beach. It is a perfect day. The beach is yours. You are lying in the sand near the water. You feel the sand molding to your body. You hear gulls cawing and the sound of surf. The breeze wafts fragrances of tide and sea air. You taste saltwater and sunshine on your lips. As you lie there, deeply relaxed, your mind and senses focused on the intense pleasure of this moment, sun comfortably heating your skin, a gentle wave washes over you.

Stay there as long as you like. Then come to the page, set your timer for five minutes, and write your inner experience—the sensory feelings, the experience of the wave, the emotional context, and anything else. Stay curious and follow the pen or keyboard.

Reflection

What do you notice? Any surprises? What was your experience of the writing itself—how did it feel to be writing it? Any thoughts about neurons that might have been firing and wiring together? Make your observations here.

Three Feeling Words

_____

_____

_____

Conversations

Ann: While mentally moving from water to emotions, I started wondering about substituting other liquids for the water. How would the path be different if I poured cream or something more viscous like honey or syrup? What if I poured an acidic liquid—vinegar or lemon juice? Would that affect the pile of sand differently? Then I wondered how the emotional distance or closeness between a person who instigates an emotion and the one who feels it affects the new paths.

Deborah responds: Your wandering thoughts took you into the realm of attachment theory. When we're born we have circuitry (experience-expectant) that already has been wired to help us use sensory input to create sensory experience, such as light transmitting to sight, sound to hearing, etc.

The wash of emotion is part of the experience-dependent wiring that happens and forms the basis for how we attach to those who care for us. That circuitry is different for the different primary caregivers in our lives. So to extend your analogy we may have circuitry that was wired because of honey experiences, vinegar experiences, etc. We don't have a monolithic attachment style; our attachments are dependent on the people to whom we attached. Therefore, we each have varied styles and circuitry.

I wish all young parents knew the downside to multi-tasking during infant feedings. When parents text, tweet, catch up on Facebook, or even read a book or a work report when they are feeding their babies, they are depriving their infant of vital engagement. The feeding distance is the initial optimal distance for establishing eye contact with an infant. Eye contact, and its accompanying sensory communication such as smiles, coos, tickles, and pats, is a primary way that infants bond to their caregivers. When the parent is distracted by devices and does not make sustained eye contact with the infant, the bonding that comes from reading faces and washes of emotion is diluted.

Vivien: I tried to think of the ways that neurons have strengthened relationships and fired together in my own world, and I found myself reflecting on the inner negative feedback loop that I discovered in the Velcro/Teflon writes.

I've been thinking about that a lot and talking about it with other people, and I hope I haven't Velcroed it to the point of permanency!

Deborah responds: To reassure you, you're not adding to the stickiness—at least, not about the negative feedback loop itself. Noticing the old pattern, talking about it with others, writing about it in your journal, or simply reflecting on it invites a different relationship with that pattern. When you come to it from a place of curiosity, or when you are exercising your observational muscle, like you do in a reflection write, you're not steeped in the middle of re-firing that old circuitry. You're observing it. That's different. With time, focus and attention, you can build new circuitry that doesn't exactly negate the old pattern but rather scoots it out of the way and takes its place in foreground.

Damaris: I remember the many times I tried to stop smoking. I finally succeeded when I took up physical exercise to fight the cravings. Every time I got an urge to smoke a cigarette, I did something to get my heart rate up. First thing in the morning was the worst time. So I took up exercising first thing in the morning. Thirty-two years later, I still exercise in the morning, and I still don't smoke.

Deborah responds: Your embodiment story reminds me of the current evidence from the world of trauma treatment that demonstrates that talk therapy by itself isn't sufficient for healing. Healing requires embodiment of some sort; trauma is embodied and the resolution must also be embodied. Your exercise solution is a perfect example.

Writing is a way of embodiment, particularly if the writing is done by hand and there is direct connection with the page. I've noticed that even when I'm not in a situation where I can write about something, simply holding the pen has a calming effect on my nervous system, particularly if it's one of the pens that I get from the yoga studio where I currently take classes.

Lee: Does that mean I have to write by hand? My handwriting is terrible and nobody can read it. Besides, if I wrote on my laptop or tablet I could password-protect it.

Deborah responds: It's better to write on a keyboard than to not write at all, and your intention, attention, and action are collectively more important than your choice of writing methods. However, the brain gets more of a work-out when you are writing by hand. It's like the little incidental pieces of exercise you get during the day: climbing the stairs in your three-level house, parking your car at the far end of the lot—you don't think of them as exercise, but when they are part of your routine, they add up to healthier habits.

With keyboarding, the brain is not asked to make distinctions between formation of letters—they're all the same—and while there is an embodied experience, it is not of the same scope. However, in the end, intentional writing is intentional writing, and the most important thing is to do it.

Kay responds: If nobody can read your handwriting, that's sort of like password protection, isn't it?

I did a study about digital journaling (Adams, Ohren, and Hudson 2015), and one of the interesting findings was how men perceive handwriting and keyboarding.

Fifty-nine percent of men and 48 percent of women said that they got the same or similar outcomes whether they wrote by hand or by keyboard. And 74 percent of men and seventy-three percent of women expressed that they were comfortable typing. However, when asked the same question about writing by hand, only 56 percent of men said they were comfortable! (At 76 percent, women were even more comfortable writing by hand than by keyboard.)

This eighteen-point gap in comfort level for men between keyboarding and handwriting may well represent guys who might be inclined not to write at all if they were required to write by hand. And that gets my attention. I would much rather have men (or women) write on a keyboard if that's what will get and keep them engaged than to struggle with comfort level or to be unwilling to write at all.

The Words You Speak

The words you speak become the house you live in.

—*Hafiz*

I come from a family of worriers, particularly my mother. And I've wondered what experiences caused her to be this way. Perhaps part of it was from a trauma her family suffered when she was a teen. It could be genetic, or perhaps my grandmother was a champion worrier too. It could and probably did result from a variety of causes.

I have inherited this trait but in recent years have become aware that I can change it and be more at peace. I have learned to trust life and the universe. If something happens, it happens. Truly, by far, most of what I have worried about never came true. And the more I remember that, the stronger the feeling of trust becomes. I did not realize that I was consciously creating new neural pathways as I did this.

—*Hannah*

I'm intrigued to think how consciously creating positive, calming experiences and feelings—and associating them with other experiences and senses—could create new linkages I can consciously evoke to reduce stress and promote my own well-being.

—*Maggie*

The sand analogy has helped me to see these concepts more clearly. There are deeply entrenched sand patterns, awash with emotion, from previous experiences, but I don't have to continue to deepen the trench. I can take my attention elsewhere to a fresh site and groove it.

—*Emily*

12

For Your Olfactory Delight

DEBORAH ROSS

Smell is the most primitive, immediately direct of our senses, particularly when it comes to activating our long-term memory centers. In its role as gatekeeper for nostalgia we can be transported to Grannie's kitchen, the summers at the lake, or the intensity of first romance, complete with images and emotions. These sensations are evoked so quickly by scent that our cognitively based editing system hasn't even booted up before we are transported.

For these reasons, the sense of smell is one of the fastest and most reliable ways to get the brain to wake up and take note. Therefore, we'll pair a scent with many of our writes directed at positive change.

The first task is to select something with a scent that appeals to you that can accompany you as you recall and write. My personal favorite is lavender, because it is both energizing and calming. I buy my lavender from a farmer who mindfully nurtures all his crops, and I think the absorbed love is also part of my inhaled experience.

Choose a scent that has a personal significance to you and that you associate with one or more qualities you value (e.g. well-being, contentment, love). Start in your kitchen (herbs, spices, extracts, citrus) or bath (soaps, shampoos, lotions, cosmetics, candles) or in nature (beach, rain, forest, garden). Choose a scent that you can use over the next several processes, one that is actually available to you (e.g., a lemon, a bag of herbs, a spice bottle, an essential oil, a cologne). For this series you will need to have the scent with you as you recall and write, not just the memory of it, although later you can use the memory of a scent if the actual scent isn't accessible.

It's fine to have more than one scent as you move through the Inklings, but for now just choose one that is accessible to you.

I invite you to use your selected scent to add another element to the process of recall and writing. Kay will guide you through the Captured Moment technique. Inhale your scent as you notice the moment you wish to capture, allow it to help you set the intention for recall, catch whiffs while you write. Most importantly, use it to help you write your reflection. Notice what happens when scent is added to recall and writing.

From a neurological standpoint, we're considering the Teflon aspects of our brain and looking to deepen the installation of a positive experience, using the Captured Moment writing technique paired with positive olfactory association.

Kay Adams

Got your scent? Good. Take a deep whiff. You're going to write about a time you want to savor more fully, celebrate more completely, acknowledge more deeply, or pause to preserve—a moment that represents something of value to you. It may or may not be evoked by the scent you've chosen; they're actually independent variables, but if they overlap, that's fine. When you're done, read and reflect.

The technique we'll use is called captured moments. It's the prose equivalent of a camera shutter, a written vignette of a moment in time, written from the senses, allowing you to savor and preserve a particular moment that holds something of worth to you. Because the technique lands in the middle rungs of the Journal Ladder, it invites creative expression. Thus, some folks use captured moments as a place to experiment with a creative writing style, such as lavish language, emotional resonance, or evocative understatement. This time, just allow yourself to focus on savoring, appreciating, preserving—all while breathing in your scent.

Inklings

Savoring, Appreciating, Preserving

Instructions: Inhale your scent and bring to mind a time that you want to savor, appreciate, and preserve. Bring to mind the details of that time, the images, tastes, textures, scents and sounds. Remember the feelings associated with this time. Stay there as long as you wish, inhaling your scent, and when you are ready, capture the moment in prose. Set your timer for seven minutes. Continue to inhale your scent as you write. Then read and reflect.

Three Feeling Words

_____

_____

_____

Reflection

Instructions: Read what you've written, inhaling your scent as you do, and give yourself feedback. Were there surprises? What did you notice about the scent you chose? What was the process when you combined the scent with the story?

Three Feeling Words

_____

_____

CHAPTER 12

Conversations

Michael: I was surprised how immediately the images, emotions and bodily sensations emerged. It amazes me how quickly inhaling the scent of Old Spice deodorant jump-started a childhood moment with my precious father.

I was transported immediately to the image of watching my father shave in the upstairs (only) bathroom. My father took his time shaving, carefully gliding the blade across his Barbersol-cream-covered face. With each slow, smooth path of his Schick razor, his soft, clean face emerged.

This face, my dad's face, was and will always be the face of a hero, my hero. This face saw way too much desperation, devastation and death. But there he stood, my hero. That's how I will always see him, but I always wondered who he saw in the mirror. Who did he really see? A hero or just someone who survived the indescribable—polio as a child, the Great Depression, fighting at Guadalcanal in the Second World War?

Thank you, Old Spice scent. You make me feel warmer and stronger inside!

Deborah responds: A lot of sensory and relational experiences emerged from that whiff of Old Spice deodorant. You can see so many associations from childhood into adulthood along with the questions that emerged that reflect your curiosity about how your father saw himself. So many rich layers, possibilities and ultimately a sense of well-being as you notice how you feel in the presence of those memories. It demonstrates the point that scent deepens an experience. It's no accident that hotels often use scented toiletries and encourage you to take them with you. They want you to return.

Emily: I had a lovely time sitting in the healing garden in the retreat center I am staying in for a conference. There are a number of flowers in bloom, including honeysuckle, which is such a sweet reminder of childhood and long summer days. I wrote with its scent in the air and on my fingers.

My write was about giving birth, the three times I have done it in a physical sense and the sense of openness and vulnerability and joy and wonderment that it brought each time. The power and resourcefulness of the female human body and the absolute shamelessness and pride I felt of the whole me—emotional, physical, spiritual—in getting through it, literally surviving it. It was wonderful to twin this with the honeysuckle, an evocative smell for me, a seasonal smell, an indicator of the warmth in the air, the turning of the season; the freshness, the scent of summer, the simplicity and patience and regularity of the flower and life cycle.

I'm anticipating that the next time I encounter honeysuckle it will bring back this sense and the cascading set of feelings and warmth I have about new life. Time will tell!

I wonder—because honeysuckle is such a specific scent, does everyone have essentially the same associations to it?

Deborah responds: What power in your recollections! Mind, body, spirit all wrapped up in that scent which is a harbinger of a new season. And because it's on bushes and vines you don't need to be up close to catch a whiff. The air currents will deliver it to you.

Interesting question about shared scent memory. What I imagine is that we're all cued to easily recognizable scents, whether in nature or in the culture, all of which have associations for each of us. Given all the research on how we create neural maps, each of us undoubtedly has plenty of personal associations for familiar scents. I think your question suggests that a language cue that names a scent opens the door to the implicit memories that were stored with that sensory association (neurons that fire together wire together; scent is a particularly powerful, primitive, direct, and evocative sensory association). As a case in point, your naming of childbirth didn't particularly cue childbirth for me, but naming honeysuckle cued a memory of a hike with my oldest son before he left to teach in the Middle East several years ago.

Vivien: I chose to use Pert Plus, the shampoo, for this exercise. When I was 23, I left my job at an investment firm in DC to move to an organic farm in northern California, close to where I now live. It was a bold move for me, when I had not made many bold moves in my life, and I remember choosing to buy Pert Plus very consciously, so that I could take faster showers (that shampoo/conditioner combo really saved a lot of time and water).

When I smelled it, I was immediately transported back to a very specific, special time in my life. I was instantly back in the outdoor showers, under a grove of redwoods, each a private stall without a ceiling, and that beautiful woodsy smell. I remembered the quiet of the moment, granted by the density of those trees, and the combination of the cool air and the warm sun upon my bare skin. Amazing. And I remembered, too, the feeling of freedom that surrounded that time for me. It was the first time that I'd taken an unconventional step in my life, and the lessons of that action, and of finding myself so joyous in a place that I'd never before even visited, opened a world of possibilities for my life. That decision continues to be one of the fundamental stories of my life, and I'm so grateful for it. And so happy that I could revisit it in a bottle of shampoo! I had a hard time putting it down.

Deborah responds: From that little bottle of practicality comes a wealth of experience, and a story that sounds like it's a resource for you to this day.

13
The Masking of a Negative Is Not a Positive

DEBORAH ROSS

I ask a friend in the midst of a difficult divorce: "How are you doing?"

She replies, "Not bad, really. It could be so much worse. We're still fighting over the little stuff, but we've stopped beating each other up over the important things. I don't think it will drag on much longer."

My friend thinks she's sharing good news. But her brain is still registering a train wreck. She is reinforcing through her language the difficulties of her current life.

Your brain registers any negative reference at face value. It doesn't hear the *not* in *not bad* or the *stopped* in *stopped beating each other up*. As far as your brain is concerned, the absence of a negative is not a positive. The brain treats even masked or diluted negativity as full-on, straight-shooting negativity. Thus, my friend's brain has translated her relatively upbeat news about her divorce as:

"bad . . . worse . . . fighting . . . beating each other up . . . drag on. . . ."

We hear people set their goals as masked negatives all the time: *I want to lose weight. I want to stop yelling at my kids. I want to be less judgmental of my co-worker.* Masked negatives also exist in the language of denial: "I'm not bad . . . It could be so much worse. . . .

My friend thinks she's made a hopeful statement, and my brain was able to translate it so that I got the general message: she's getting along. Her brain? It registered another dismal day in a bad marriage.

If you want to check this out, go to any coffee shop with your journal and eavesdrop. Write down random snatches of conversation and notice how often someone actually says, "I feel relaxed" (vs. not so stressed) or "My knee feels normal today" (vs. doesn't hurt much today) or "I am happy" (vs. I am not so sad).

When you hear yourself say, "I'm feeling less _____," or "I don't want _____," pause to consider what you are feeling more of or what you do want, and say that instead. "I don't want to be sick," is not the same as saying, "I want to be healthy." Say them both out loud and notice the difference in your body!

Kay Adams

Put your (helpful, not critical) Inner Editor to work and catch those masked negatives when you write them. Cross them out and rewrite them as true positives. Read them out loud several times and take note of how it feels to re-language your experience. What is it like to move in the direction of what you do want, rather than moving away from what you don't want? Your journal is your field notebook as you perform an experiment of one.

When you cross out your masked negatives, don't scribble or mark harshly. Each simple strike-out followed by thoughtful rewrite gives a visual snapshot of positive change: no shame, no blame. Just cross it out, think it through, choose different words, and move on. You're teaching yourself that it's simple and natural to rewind and rewrite over, around, and through the negative.

There are two Inklings followed by a reflection. Plan some time—fifteen or twenty minutes—for the set. They're short, but they require attention.

Deborah Ross

The writing processes that follow are invitations for noticing and exploring. As Kay said, this is not a VIP invitation to your Inner Editor to get critical or smug. It is an invitation to expand, deepen and enrich the noticing of the positive.

Inklings

Finding Positive Language

Instructions: Rewrite these statements to reflect positive language only.

Example: I want to feel less anxious | I want to feel calm, grounded and relaxed.

Three Feeling Words

_____
_____
_____

1. (How are you doing?) Not bad, really. It could be so much worse. We're still fighting over the little stuff, but we've stopped beating each other up over the important things. I don't think it will drag on much longer.

2. I want to be stop being so judgmental of my oldest son when he underachieves.

3. Write your own masked negative statement, then rewrite it.

Rewriting Your Negatives

Instructions: Write a typical paragraph in your regular voice about the current reality of something that is frustrating or mildly difficult for you—a work project, a communication impasse, a household matter that isn't resolving on its own. When you're done, go back and search for any masked negatives or qualified comparisons. Strike them out and thoughtfully pause to consider what you really mean. Then re-language in the positive. Here's an example.

Original: I'm stumped about how to proceed with my business plan. I'm not as confused as I once was, but I'm still stuck about how to prioritize. I'm frustrated with the slow pace of it—I keep stalling out.

Revised: I'm ~~stumped about~~ still seeking solutions for how to proceed with my business plan. I'm ~~not as confused as I once was~~, better able to articulate the problems, ~~but~~ and I'm still ~~stuck about~~ exploring how to prioritize. I'm ~~frustrated~~ becoming more patient with the slow pace of it—I keep ~~stalling out~~ pausing to think it through.

Rewritten: I'm still seeking solutions for how to proceed with my business plan. I'm better able to articulate the problems, and I'm still exploring how to prioritize. I'm becoming more patient with the slow pace of it—I keep pausing to think it through.

Reflection

What do you notice about your masked negatives? How easy or difficult was it to re-language them as positives? What is it like to move in the direction of what you do want, rather than moving away from what you don't want? Read what you've written and reflect.

Three Feeling Words

_____
_____
_____

CHAPTER 13

Conversations

Lee: My Inner Skeptic wants to know: Isn't this emotional manipulation? Aren't we just throwing glitter on a skunk?

Deborah responds: I can understand the skepticism, but the answer is no, it isn't emotional manipulation. The fact is that the "negative and difficult" gets well over its fair share of focus and air time. Searching out and intentionally rewriting the negative into positive statements increases the bandwidth for the positive experiences that we've already had or anticipate having.

Kay responds: You do raise a good point, that every emotion deserves air time and expression, and there are going to be plenty of times when it would be cumbersome at best and inauthentic at worst to try to language genuine challenges in the positive. Note, though, that Deborah's friend was talking about something that actually did feel encouraging and hopeful for her. I think the intention/attention/action is at play here; let's pretend her friend, we'll call her Flo, is reading this book and rewriting her story.

> *Intention:* Flo wants to change the way she languages her divorce.
>
> *Attention:* Flo listens to herself and hears what she says about her divorce. She notices the places where she delivers good news as masked negatives.
>
> *Action:* Flo writes down her masked negative talk and then rewrites it as positive. There's likely to be a "ping" in there, a word or phrase that sticks. Flo might do a seven-minute write—a sprint or a cluster—using that word or phrase as a prompt. Then she intentionally inserts that word/phrase in conversations.

By all means, vent and rant and be sad in your journal when that's how you're feeling—your brain may not love the language, but your journal grows and becomes more when you offer it your problems and difficulties. Deborah and I prefer that you keep that journal separately from your Brain on Ink journal while you're working with this book, but once you've learned the process, you can merge them if it works best for you.

14

From Positive State to Positive Trait

DEBORAH ROSS

Once the neurons are firing and wiring, and new circuitry is beginning, how do we keep the momentum going? How do we create a new neural path?

The answer is installation—the process by which the positive state is moved to long-term memory. Imagine that you have purchased software for your computer, but to activate it you must download and install it. The software then becomes fully integrated with the computer, the path of least resistance to prepare your spreadsheets or update your mailing list. Until that magic moment of installation, however, you will not be able to release its possibilities.

Rick Hanson (2011) has a three-stage metaphor describing how we move from positive mental state to positive neural trait. Hanson compares the installation experience to building a fire:

- Light the fire. Gather the "kindling"—your positive experiences, goals and aspirations, best relationships. Bring, too, your writing techniques and journal.
- Add logs. Once you've got the fire lit, add in your intention and attention. Enrich the experience. As we have learned, the brain makes faster and deeper associations when the positive experience is paired with sensory input and washes of emotion. This offers the action-oriented opportunities to use familiar and well-loved scents, as we did in the last chapter, and to seek and write about positive experiences that awaken emotional connection.

- Take in the warmth. Allow yourself to absorb the experience. Hold the experience in memory, and make it last through critical-mass sequences of five, ten, and twenty seconds. The twenty-second process comes from learning theory, which suggests that learning requires holding information past that short-term memory buffer.

Give your novelty-loving brain something to engage it; it will help your brain learn something new or deepen an existing positive memory. Write with a purple gel pen (Kay's favorite) or illustrate your write with a crayon border. Go to a coffee shop you've never been to before or write in the park. You can take novelty even further by identifying the personal relevance: How does this novelty choice relate to me? Why is this important? How does it feed my soul?

As you have learned, the brain is very capable of nurturing the negative. It doesn't completely ignore the positive, but the standard operating equipment of your brain is designed to help keep you reasonably safe in the world by reacting and responding to perceived threats. This is a good thing! Yet if you want that protection and a deeper, richer, more harmonious, more resilient life, self-directed neuroplasticity requires that you be intentional, attentive, and action-oriented about it.

Installation is the key to moving from positive state to positive trait.

Kay Adams

It's a Saturday afternoon in 1967. The Rathskeller, a German pub in the freshly hip Larimer Square in downtown Denver, has been invaded by twenty student journalists from high school newspaper staffs, piled like puppies around three tables. Each is electric from a just-adjourned press conference with the supermodel Twiggy on her first American tour. Everyone orders coffee and cheesecake. It is my first cheesecake. It tastes exotic and creamy. It tastes like risk and success. I am awash with the joy of being a proud writer and editor among cohorts who share my passion.

Over the years my first cheesecake has become a signature story for me. I have installed it as an early imprint of success and a snapshot of a moment that foreshadowed my career: I'm still writing, interviewing, taking risks, loving my work, loving my colleagues, celebrating with cheesecake.

To deconstruct this through the installation metaphor: My fire got lit with the positive experience of my first cheesecake and all that became inexorably linked to it. I no longer have the Big Chief newsprint tablets that served as my journals in high school, but I can guarantee you that the experience was captured there, written on the page and on my bones.

The logs to stoke the fire are added each time I remember and write the cheesecake story, which I've done dozens of times, always with intention and attention. I observe myself in the telling, always asking, What does this story teach me today? Each time I feel the familiar wash of emotion: exhilaration, celebration, success, joy, connection, the thrill of risk, community. Each time it is inseparable from the cheesecake: I feel the creamy texture against the roof of my mouth, I taste the buttery crumb crust, I inhale the rich aroma of coffee in a thick blue mug.

I take in the warmth through repeated harvesting of wisdom from the experience. Over time it has become a defining story of my particular imprint of success, offering up a recipe: equal parts passion, talent, skill, hard work, friends, and smart risk.

Sometimes I write this story in writing group, and I share it with others who witness with respect and attention. I have written this story in my own therapy and shared it with my therapist. Usually I write in solitude and read the story out loud to myself, in different voices. Each of these ways is also powerful enrichment.

In this next write you will return to the captured moment you wrote about the olfactory experience. To prepare, you'll need your scent cue, your previous captured moment, and a timer that keeps seconds (a quick internet search yields choices). Set the timer to twenty seconds but don't start it yet.

To add novelty, also choose a different pen or pencil, such as one with brightly colored ink/lead, or add music, or change the location where you write.

Breathing in your scent, read the previous captured moment, start the seconds timer, and hold your story consciously for at least twenty seconds.

When you are ready, proceed to the next write.

CHAPTER 14

Inklings

Captured Moment 2: Going Deeper, Seeking Meaning

Instructions: Starting with the same event or moment that you wrote about in chapter 12, go deeper. Place your intention and attention on this story and focus particularly on its meaning: What are the take-aways for you? What is the learning offered you in this experience? How can you apply action that takes steps toward your own healing, growth, and change? When you are finished, read and continue with the reflection write.

Three Feeling Words

_____
_____
_____

Reflection

Instructions: Write a reflection, paying particular attention to how this story may be changing or deepening, and what meaning you may be deriving from it.

Three Feeling Words

_____
_____
_____

Conversations

Charlotte: I'm planning to do my best with this activity, but I have a very physical challenge here: I have almost no sense of smell. I rarely smell anything and on the rare occasions I become aware of a scent, it isn't strong enough for me distinguish what it is. I apparently inherited this from my paternal grandmother who also had minimal sense of smell. One scent that I have slightly better awareness of is vanilla, so I'll try using vanilla for this activity and hopefully it will work.

Deborah responds: You can try the vanilla and if it works that's great. But I'm wondering whether there is another sense—sound?—that might work better for you? Or maybe something soft to invoke touch? When you call up positive memories do you notice a particular sense that seems most active? Please just go with what you think might work best for you.

Emily: I finished my write before I left the retreat center. I picked some wild honeysuckle from the vine outside my window—the sweet, sweet smell that takes me back to childhood—bringing it up to my nose and taking a long inhaling breath while sitting in a natural enclosure called the Healing Garden. What a lovely experience and metaphor to take with me! And it was a novelty aspect. I don't normally journal outside and certainly haven't journaled in something that is specifically set up as a healing garden with all that that connotes. I noticed that the act of entering the garden through its archway almost unconsciously set an intention to make this a healing or rewarding experience.

What I found was the smell was cleaner and fresher and clearer outside, and that seemed to translate to the memories. They also had clearer edges.

Deborah responds: Here is the play between intention and attention, or noticing. I imagine you are priming the pump or setting the stage for a more focused, less diffuse experience, hence a fresher, clearer experience. I do a lot of my journaling inside but have found that when I link writing to an outside experience—going to the park, walking an outdoor labyrinth, even sitting on my own porch—it adds a powerful level of sensory input that isn't there when I'm writing indoors, even with the window open.

Hannah: I still used a lemon scent, but this time chose an experience unrelated to it: riding with my husband on his BMW touring motorcycle when we lived in the southwest. Specifically, going out early on Sunday mornings to breakfast in nearby towns. It was a highly physical experience, riding on a motorcycle and not in the box of a car, and, as I realized in my reflection, a kind of intimate sharing that we no longer have since his accident and brain injury and (thank goodness!) decision to quit riding.

It was interesting using the lemon fragrance to install the memory of this moment more deeply. Perhaps the delightful "shock" of the lemon heightened my recall of other scents on those rides, like the pines as we rode upward from the desert floor into the mountains, and the feel of the air temperature changing from HOT to wonderfully cool.

I also wrote with a pencil and wrote across the long side of the paper instead of the usual shorter side. I noticed the scratch of the pencil (I usually use a smooth-writing pen) and how my arm moved differently to accommodate the length of the page. Perhaps those elements of this write increased my recall of those rides years ago. In any case, I am intrigued by this technique and will use it more.

Deborah responds: The lemon, the pencil, the long side of the paper—three novel cues to nudge your brain into installing or deepening memory and recall.

Shannon: I think I was most struck by the deepening aspect of this process—how the same stimulus (scent) paired with a holding of the experience and writing in a different way/place (this time a different geographic location) was enriching and felt both familiar and new. I smell this scent often and wear it almost daily, but the invocation to have such intention around it deepened the process and engaged even more of my senses in the writing. Another surprise was the juxtaposition of presence and longing, and the awareness that they could be present at the same time. I didn't notice until now the feeling of expansion in my heart and solar plexus after the installation, something not present before the deepening of the experience through repetition and holding.

Deborah responds: Yes, you noticed the embodied sense of expansion. You were inviting multiple modes of engagement—scent, novelty, focused attention—and the combination seemed to reinforce itself. As you develop or strengthen the observational muscle, you can pick up juxtapositions or multiple ways of experiencing. It's all about noticing the vista of our experience.

15
Building Neural Circuitry

DEBORAH ROSS

Kornfeld (2008) says mindfulness is "patient, receptive, non-judging awareness." Poon and Danoff-Burg (2011) say it is "paying complete attention to the experiences occurring presently, in a nonjudgmental way or an accepting stance." My meditation teacher defines mindfulness as "remembering to pay attention." Focus, mindfulness, and attention are three keys to building new neural circuitry. A fourth key is a natural outcome of the others: the older, less used circuits will fade as a result of the increased attention in a different direction. Although old circuitry doesn't seem to disappear entirely, at least not in adulthood, it can be superseded by new circuitry.

I liken the process to planting an organic garden. The gardener considers the plants he wishes to cultivate and then uses all the health-building strategies he knows to help the plants thrive. In the process, the thriving plants crowd out the weeds. It is similar to neural circuitry: the robust health of one creates conditions that lead to the diminishment of the other.

The brain begins a pruning process in infancy and carries it through until the mid-twenties. Circuits that aren't used much are pruned out; this allows more frequently used circuits to become strengthened. If a child is abused or neglected and her empathy circuitry is quite weak, it will likely be pruned; this subsequently reinforces disconnection and problematic behaviors.

The literature on attachment is informed by the idea that a wash of emotion contributes to, or is even largely responsible for, the wiring of our neurons into circuits, beginning in infancy. As soon as we are born, our circuitry starts developing, firing, and wiring according to the quality of our interpersonal relationships with (hopefully loving) caregivers.

Add one more piece to this puzzle: there does appear to be an optimal distance for that initial wiring. That distance is the eye-gaze distance from baby to parent or caregiver when the infant is nursing or bottle feeding.

How does all this apply to the self-directed neuroplasticity through writing?

We have emotionally wired circuits that can be modified by how and where we place attention. Hence the importance of the reflection write; it deepens our experience of noticing what is happening as we write, and it often yields suggestions for the next piece of writing, or offers ideas for action steps to take.

Writes that are accompanied by an emotional undertow seem to be more powerful than those that are more a recitation of facts. In the classic expressive writing research model developed by Dr. James Pennebaker (2004), participants who wrote about both facts and feelings experienced longer lasting benefits than those who wrote only one or the other.

From the world of epilepsy comes the notion that seizures beget seizures—each time a seizure happens, another small group of adjacent neurons is recruited and is now more likely to fire with the first group the next time around. Perhaps writing from any emotional place might also be recruiting additional small bands of adjacent neurons that will fire with the first band. In other words, the more you express resentment in your journal, perhaps the more likely you are to experience and express resentment—or, conversely, expression of positivity may beget more positivity in your writing and thinking.

Kay and I were discussing the gaze distance research and noticing how holding a feeding baby while sitting tends to give a particular lean in the body. Kay observed that it seems similar to how we are positioned when we write by hand. Might the act of writing our deep feelings by hand connect us more deeply with ourselves, just as we originally became more connected to those who cared for us? Writing can be seen as a significant act of self-nurturing. The practice of mindful writing can lay the track for new neural circuitry.

Conversations

Vivien: I love these questions, Deborah, and am really reflecting on the posture question, especially having just spent a lot of time around my sister as she nurses my nephew. I've always really hunched down when I write by hand, and there have been times in my life when I've noticed how very physical the act of writing can be. In writing emotion-laden pieces, I curl up over my paper even more, so much so that sometimes I recall a line from a college reading: *I am a woman giving birth to myself.* And sometimes with writing it is exactly that.

Deborah responds: What a powerful observation! Continue to pay attention.

Kay responds: As you pay attention, Vivien, you may want to take notes. When I'm in simultaneous process and observation mode, I like to leave the left page of my notebook or journal blank and write only on the right side. Switch it up if you're left-handed. Thus I have a "parking lot" for notes on process that I might come back to for the reflection—or I may just leave them as jottings, field notes that accompany the experiment.

16
Series of Three

KAY ADAMS

Continuing the theme of installation—the process by which positive memories and experiences become more Velcro-like—we come to the series of three.

The series of three uses sequential writes, either using the same technique or using three different techniques. Each write builds on, deepens, or extends the one before it. The layering fosters installation because you are focusing on the same material from multiple angles in a short span of time, and the choice of the material will be personally meaningful in a positive way.

We'll use the familiar captured moments technique, followed by a character sketch and then by a shift in perspectives.

A character sketch is a written portrait of yourself, or some aspect of self, or of another person. It often begins with a physical description—the exterior self presented to the world—followed by a deepening into the internal character, goals and aspirations, motivations, and qualities that become the lens through which this self or aspect of self views the world. When you write a character sketch about a time when you felt successful (as you are about to do in this series of three), you have the opportunity to discern individual components of your particular success pattern, the qualities you can intentionally call forth and place attention on as you move into action-orientation.

The character sketch technique is in the middle rungs of the Journal Ladder, indicating a propensity toward insight and a flair for the creative or imaginal.

The perspectives technique, which you'll use for the third Inkling, is in the upper rungs of the Journal Ladder. It is characterized by an intentional stepping away from concrete reality and the entering into a different time/space/voice continuum. There are two basic ways to shift perspectives. One is to change time or space: write as if it is the future or the past (actually date your page with the month and year of your shifted perspective), or write as if it is current or altered time and you are in a different geographic place. The second way is to change voice: write in the third person (the s/he voice) about your own life situation, or assume the persona of someone else or some aspect of yourself, and write in the first-person voice from that point of view. The novelty of this altered perspective can bring significant clarity.

Following are three Inklings. Do them sequentially. Each should take about seven or eight minutes; with the reflection, you're looking at about a twenty-five to thirty-minute writing time. Use your scent cue if you wish to help with installation.

Inklings

First Write: A Time of Success

Three Feeling Words

Instructions: Bring to mind a time when you had an experience of inside-out success—a time when you wanted something and set your *intention*, focused your *attention*, and went into *action* to make it happen (even though the articulation or recognition of these steps may not have been conscious). Be sure it is an experience that you wanted, independent of what others wanted for you, or what you thought you *should* want but actually didn't care about that much.

When you have the experience, bring it to mind and focus on it for twenty seconds. If you're using a scent cue, inhale deeply. Then write a captured moment of that time in the present tense as if it were happening right now, focusing on sensory details. Write for seven or eight minutes by the timer, then move on to the second Inkling.

Second Write: Character Sketch of Successful Self

Instruction: Bring to mind an image of yourself in that moment of success. Notice first how you look, what you are wearing, your facial expression, the aura you project. Then turn inside and explore the feelings. What does it mean to be having this experience of success? Which of your core values does it reflect? What does it represent to you? What is the meaning of this success? How does it shape what you perceive as possible for yourself? Hold the image for at least twenty seconds, using your scent cue if you wish. Then write about yourself in the third person (the "s/he" voice), as if you are a compassionate narrator. Again, write for seven to eight minutes by the timer, then move on to the third Inkling.

Third Write: A Future Success

Instructions: Bring to mind a project, process, or decision that has inside-out value for you—something you genuinely care about because it lights you up inside, invokes your passion, or has meaning for you. Project yourself a few months into the future, and date your page accordingly. Now imagine that you have brought your successful self from the second write into this process, and that aspect of yourself has been in charge. What is the outcome? What has the process been like? What has worked particularly well? How did your successful self overcome obstacles or resolve difficulties? Hold the scene for at least twenty seconds, using your scent cue if you wish, and then write for seven or eight minutes, in the first person, present tense. When you are complete, read all three writes and continue to the reflection.

Reflection

Reflect on the three-stage process. What surprised you? Where did you feel this in your body? What can you discern that is useful going forward?

Three Feeling Words

_____

_____

_____

17
And Now For Something Completely Different

I am large. I contain multitudes.

—*Whitman (1892)*

DEBORAH ROSS

We've been focused so far on the need to encourage more Velcro for the positive. We've worked with noticing the positive, taking it in, deepening the experience by cuing it with sensory input such as scent, including novelty in our writing, and finding personal relevance and meaning. All of this reinforces the positive and helps retrain the brain.

But what about the part of you that is a little squirrely about all this positivity? The part that may be more at home in the Velcro world of negativity and may be more accustomed to the sharp inhale than the long languid exhale?

Well, this chapter is for that part of you.

The part of you that may be skeptical, nervous, or dismissive deserves a voice, too. It is part of the totality of your entire being, one of the multiplicity of "parts" in the internal system.

So let's provide some air time for any parts of you that might be skeptical about this big dose of positivity.

It's important that you show up with curiosity and compassion for any parts that struggle with change, especially positive change. Bringing these concerns to the page where they can be articulated and processed with a teamwork approach keeps them conscious. Otherwise our efforts might be working at cross purposes at the unconscious level.

There are various branches of psychotherapy that address what Italian psychiatrist Roberto Assagioli (2000) called subpersonalities (aspects of self): Gestalt, psychodrama, Virginia Satir's work with family therapy, Hal and Sidra Stone's Voice Dialogue, Carl Jung's concepts of archetypes, and the one I am most informed by, Internal Family Systems.[1] Each is organized around the same principle: When we acknowledge and engage the multiplicity of our complex personality systems, we can forge alliances with our different "parts," give voice to each, and find creative, holistic solutions to shift internal discomfort or conflict.

Kay Adams

Dr. Ira Progoff, whom I consider to be the founder of the contemporary journal therapy movement, devotes six chapters of his classic *At a Journal Workshop* (1992) to the various dialogues one might have in one's journal. Although he does not specifically suggest dialoguing with subpersonalities or "parts," he advocates for both the dialogue with persons (including abstractions such as fictional characters or persons who have not yet materialized in one's life—an unborn child, a future mate), and the dialogue with symbolic energy (including representations of inner wisdom, including inner archetypes).

The true Progoff dialogue, facilitated by one trained in the method, is rich in ritual and depth.[2] The more accessible layperson's version begins with a brief guided imagery meditation that includes agreement on two points:

1. Either party may ask any question and receive an honest answer (even if the answer is *I don't know* or *I don't want to say*), and
2. Either party may make any statement, knowing that it will be received with thoughtful respect.

With those guidelines in place, a curious conversation unfolds like a movie or play script:

Me: Skeptical part, will you speak with me?

S. P.: Hmmm, sure, I guess so.

Me: What do you think about all this positivity? Does it seem crazy to you?

S. P.: Well, sort of. It's so pie in the sky, loosey goosey, snake oil. You know?

Me: Sometimes it sounds too good to be true. Can you tell me more about how it comes across to you? Is there something in particular you don't want to let go of?

The dialogue technique is one of the Journal Ladder's upper-rung techniques for several reasons. Although the back and forth of the written dialogue provides a strong structure, the technique is served by a spacious time boundary—at least twenty minutes—and the capacity to hold a deeply receptive internal space. The writing does not always flow. Sometimes there is attentive silence between the exchanges; then words bubble up, or there is a sense of taking dictation from an internal source with a concomitant obedience to the task of faithfully transcribing thoughts and words.

Write what comes. If nothing comes, sit quietly and listen. If you feel off-track, ask a different question or rephrase your statement. Sometimes it is helpful to physically change positions by sitting in a different chair or swiveling to a different part of the desk or table. Writing a character sketch of your dialogue partner as a warm-up can be a way of easing into the dialogue.

In the first part of this write, you'll identify your dialogue partner—some aspect of yourself that is not in alignment with all this positivity—and you'll give it a descriptive name, such as the Skeptical Part. Get as creative with your naming as you wish.

Then you'll write a character sketch of this "part." Let your creativity out to play and remember that details help bring the character alive.

In the second write, you'll dialogue with this aspect. A good opening dialogue can be had in fifteen to twenty minutes. If you're going strong, go longer. Remember to bring open curiosity and an attitude of active listening and respect on both sides of the dialogue.

As you bring your dialogue to a close, it's a nice touch to thank your partner and ask if you can talk again.

If you've never done a written dialogue before, you may be surprised at what emerges. People often report that they "don't know where that came from," and they express surprise at the clarity, wisdom, and "write on" insights that this metaphoric communication offers. It's fine to be temporarily uncomfortable. Unless you're really bothered (in which case, stop the dialogue and shift into the reflection write), see if you can roll with it and observe with curiosity and engagement. When finished with the dialogue, read both writes and move on to the reflection.

Inklings

Character Sketch of a Skeptical Part

Instructions: Close your eyes and allow a name to come to you that describes the part of you unwilling or unready to give up negativity. When you have a name, allow a mental image of this aspect of yourself to form and notice details. Also note emotional and personality characteristics, motivations and urges: What does this part want to do, be, or have? What does it want from you? What motivates this part? When you are ready, set the timer for ten minutes and write a character sketch. Upon completion, proceed to the dialogue write.

Three Feeling Words

_____

_____

_____

Dialogue with a Skeptical Part

Instructions: Bringing forward the character you just created, close your eyes and take some deep breaths. Imagine that you are in a safe place in nature, a place you enjoy. You're on the way to a meeting with your dialogue partner. Check in with your feelings. Gather your thoughts and questions. See your dialogue partner in the distance, approaching you. As you join, make an agreement that either can ask any question and receive a truthful answer, and either can make any statement and be heard with compassionate ears. Set the timer for fifteen minutes; extend if you aren't ready to quit. Write your dialogue like a script on the page. If it is helpful to change chairs, positions, or pens when you change voices, do that. Remember to exit gracefully by thanking your partner and asking if you can talk again. When complete, read and move to the reflection write.

Reflection

What did you notice about the character sketch and dialogue? Were there surprises? Insights? What was the process of writing in two voices like? What will you take away from this experience?

Three Feeling Words

_____
_____
_____

Conversations

Bridget: My write channeled the part of me that I easily recognize in my oldest brother, that I also know exists in myself. At its simplest it goes like this: If you expect the worst, you won't be disappointed. For a long time I lived like this to some extent, expecting very little for myself. I have deliberately tried to get rid of this thinking in the last ten years because it just doesn't serve me well. But, like Velcro, it can hang around.

In the dialogue with this pessimistic part of me, I was gentle but firm that this is no longer what I want. It is okay to be somewhat skeptical but not this kind of utterly negative stance. Overall, it was a rather painful examination, but for now I am intending to sit with it and revise it later to see what feelings come up.

Deborah: "At its simplest it goes like this: If you expect the worst you won't be disappointed." Many of us have a version of this part, and we often learned it from someone close by as we were growing up. Our parts don't appreciate being silenced and can come back with a louder voice if we try to do that. So gentle and firm is a good strategy, certainly better than polarization. When you can show up with curiosity and compassion, it's more invitational and often results in learning more.

Sara: My initial response in reading this week's introduction was a sense of enthusiasm. I think this was related to how positivity is often seen as a solution in and of itself—"just think positive." I don't feel this has been the tone or message of this material, however. Harnessing the power of focusing attention on positive events does not necessarily mean denial or exclusion of the neutral or negative.

I am familiar with IFS on only a superficial level but have adopted the concept of parts in my clinical work. For me, it has served to be a validation of our multiple facets and a means of not having aspects of our experience overvalued or necessarily seen as representative of our whole self. I believe this model along with mindfulness and non-judgment concepts have shifted my own conceptualization of "negative" and "positive."

This spirit was prominent in my dialogue write this week. In reading this week's prompt, I anticipated an energetic back and forth. Instead, the dialogue was mild in tone and non-conflictual. I think that when both aspects, positive and negative, can be accepted there does not need to be an either/or. Instead of a but, there can be an and.

Deborah responds: I find so much more healing in the space of *and* rather than *but* or often *or*. Polarizations diminish, and energy doesn't divert to internal struggles or, sadly, even wars. I imagine that we move through the world more gently. The longer I've been in therapeutic practice the more I'm drawn to those systems or models that come from a core of curiosity and compassion. And mindfulness seems to set it all in motion.

Hannah: My dialogue was with the part of myself that tells me I'm not good (experienced, articulate, whatever) enough to do this work I'm doing as a therapeutic writing facilitator. As I discovered, I downplay my abilities and performances out of some need to remain "humble" or not "too big for my britches." As the child of a "good Catholic family" of the 1950s and 60s, and as someone who is naturally, I think, reticent to some degree, I am often nervous about expressing how I truly feel, even to people who love and accept me. Even when what I want to express is positive and good. So my dialogue was with that part of me that fears deep self-expression to others. And I realized that over the last few years as a therapeutic writing facilitator and conference presenter, I have become more open to that kind of out-loud expression. I can let go as best I can those fears about not being "whatever" enough because evaluations and comments consistently tell me that I do just fine.

It looks like I have been retraining my brain to be more confident and outgoing without really realizing it. I haven't been worrying about these things nearly as much lately. It was just such a habit, I didn't realize that I have made tons of progress.

Deborah responds: What's so interesting to me about our wiring around humility, soft-pedaling our successes, shrinking from visibility, etc. is that all of those messages often came from the people who cared for us. These neural maps are relationally driven. We absorb the message, wire the circuitry, and it's largely to keep us in connection with the folks who presumably loved and cared for us. At this point in our lives I think that the "task" is to be in better relationship with ourselves. That's what I love about the Internal Family

Systems paradigm. If we're not waging war internally, there's more likely to be peace in our relationships.

Vivien: I was surprised by my reaction to these writes and had to take my own step back to absorb the power of allowing the negative to emerge. Dialoging with that piece was incredibly moving, in that I found a spot that I thought I had resolved. But, it turned out that inviting it in was opening the door, and it came roaring through. It's interesting; I don't think that I've falsely closed that door. It's been a couple of days since I completed the reflection write. Now it almost feels that the negativity wasn't completely real, but perhaps a strong statement by a dying presence in my life. Not that the Critic is going anywhere, of course, but maybe it is perhaps changing shape or focus. Still, it knocked me down a bit with its strength.

Deborah responds: That's the point of giving air time to those parts that don't feel that they are welcome as we focus on the positive. Sometimes an older part that has carried that energy can feel like the Wizard of Oz—a big blustering presence that is really just a person behind the curtain. Inviting it in can bring some powerful messages and some perspective.

Julia: My negativity seems to stay hidden behind words that are not spoken aloud and ideas that are not openly shared. Just connecting with these thoughts and bringing them to the forefront allows me to increase awareness and perhaps better understand the source of some frustration.

I am always working on improving my health and diet, which can be a challenge since I recognized that I'm a stress eater. I received some new insight through my dialogue noting that eating and food don't serve to restore me; instead I need to embrace my true voice. I found this meaningful, but I'm not quite sure how to relate this to brain training and if there is a way to work on both diet and the brain . . . count me in!

Deborah responds: What an important concept to examine. How might embracing and giving psychic space to that "true voice" affect the stress eating behavior? And what does that "true voice" want to say? You can find out in the character sketch and dialogue techniques—and in so doing, get some neurons firing and wiring.

As you noted, negativity doesn't have to come out in a big display but rather via the route of tamping down or hiding behind. Keep in mind that I extended a direct invitation to dialogue with a part that may feel somewhat disenfranchised in this focus on the positive. That could be why things feel a little shaded or hidden for you.

The key is the invitation. Make it intentional, a message that you are paying attention and have both willingness to listen and interest in the skeptical part's concerns. I think a big reason why we struggle the way that we do is that we spend a lot of time "shushing." One way or another, messages will be released, often with the volume amped up until we pay attention. I think it's more skillful to invite rather than deal with the inevitable clean-up later. This way, we're in charge of the circuitry that is getting wired as our neurons fire.

Notes

1. For more information on Internal Family Systems (IFS), please see the website for the Center for Self Leadership, http://www.selfleadership.org.

2. For more information on the Progoff Intensive Journal method, please see the website for Dialogue House, http://www.intensivejournal.org.

18

Anticipate a Blossoming of Creative Delight

KAY ADAMS

It was a scorching Friday in mid-July. The air conditioning in the 100-year-old psychiatric hospital wasn't working. I stepped into my journal group room and was greeted by eighteen patients glaring at me.

"Are you hot, or is something else going on?" I asked.

"We're *hot!*" they chorused.

"Well, let's get started and maybe you'll forget about it."

"We want to do something *fun* today," one patient asserted.

"Like what?" I queried, mentally comparing my facilitation plan to *fun* and coming up empty.

"We don't know. That's your job," chimed in another.

"Okay. Well. Let's write alphapoems!" I said.

Of course they didn't have any idea what an alphapoem was; it was poetic wordplay, similar to an acrostic,[1] that one of my writer friends had made up and shared with me. We had been writing them together for a few weeks, and I found them intriguing, fun, and easy.

I went to the whiteboard and wrote the alphabet, A–Z, in a column. "I'll show you," I said. "This is going to be an alphapoem about alphapoems." Then I wrote the first lines that came to me:

A nticipate a
B lossoming of
C reative
D elight!
E asy, really, once you
F ind the rhythm and the pace.

I turned from the board. "Who's got the G line?"

Silence. Nobody had the G line. I filled in a few more lines, with a little help from the patients:

G ather up the thoughts you
H old secret in your heart.
I magine them
J ust drifting out, a
K aleidoscope of
L etters
M aking words.

"See how it goes?" I said. "It's easy—no rules, remember? Who's got the N line?"

"No rules to follow," a patient offered.

"Excellent! O?"

"Well, there's the obvious rule," someone said. "You have to use the next letter, I guess."

I returned to the board. "Let's extend this line—"

N o rules to follow, except the
O bvious one.

With a little coaxing and coaching, we finished out the poem:

P erhaps you'll find a poet inside?
Q uite likely.
R ead your alphapoem; you'll find it
S tartlingly

T rue—an
U nusual way to give
V oice to your
W onderings, wanderings, whimpers, whys, wins.
X hilarating feeling to find
Y ou've reached the
Z enith of the poem!

The room broke into spontaneous applause. I looked around at eighteen damp and flushed faces, all filled with the unmistakable light of creative engagement. "Want to do another?" I asked. They nodded.

"Turn to a blank double-page in your journal and start on the left. Write the alphabet, A-Z, down the side. Use both pages if you need to. Put a title at the top: *Things I Love*. Write it like a poem, not like a list; add words to the line; extend it out," I instructed. "Ten minutes. Ready, set, go!"

Heads bowed; pens scratched. Ten minutes later, they shared their poems with each other. Much hilarity ensued; it turned out nearly everyone loved zebras and zoos, and the others were filled with zest.

We continued with another alphapoem, this time on a more serious topic: something they had a hard time talking to their therapist about. Again, they read aloud, but this time there was no laughter; each poem was a singular gem—raw, aching, haunted. After each one I asked, "Do you think you could share this poem, just as you wrote it, with your therapist?" Nearly every one said or nodded yes. Doorways to healing were opened that day.

Deborah Ross

When I first learned about alphapoems, I cringed. This was not likely to be the technique that I gravitated toward. It brought out those parts of myself that worry about not being able to find a word for all the letters and have poetry emerge.

Now it's one of my favorite techniques and a "go to" when I want to have one of those tingly integrative moments. My left brain is using the logical, linear word listing to organize my write. And my write brain is filling in the story which may include my feelings, sensations and memories.

I do prefer shorter versions that have a theme rather than simply running the alphabet. Sometimes I write an alphapoem to express an intention. I write them at the start of a new year or a new chapter in my life. I write someone's name and allow poetry to flow to express how much they mean to me. If I'm considering a new venture and want to discover the hidden poetry in it rather than just the kinds of information that I might put on a spreadsheet, I'll alphapoem the venture. And I use them in the groups that I facilitate. The last one I wrote was to say THANK YOU to a group that had invited me to talk about journaling, cancer and "chemo brain."

T o be
H ere with
A ll of you this morning
N oticing
K indnesses extended and reflections of
Y our journey, I wish you
O ngoing vitality, resilience, healing and
U nity in your support for your intentions for yourself and one another.

Kay Adams

Alphapoems are in the middle rungs of the Journal Ladder. They have a beautiful structure; they are contained by the number of letters down the spine of the poem; and the pacing is, paradoxically, at once staccato and flow. Because an alphapoem thrives on spontaneity, there's faster access to what is beneath the surface and beyond the obvious. Alphapoems can be deeply insightful and quite creative, making them an excellent choice when you're leaning into something challenging, when you want immediate insight, or when you simply want to play with words. I have deep respect for the consistency with which even staunch non-poets are able to write powerful, relevant, illuminating poems in just a few scant minutes.

In that first alphapoem group I realized that this structure, when used with intention and attention, was anything but wordplay; it was a powerful tool for insight and awareness. Soon after, I realized the form also holds for any word or phrase, an emotion or quality, a problem or challenge, even someone's name as the spine of an alphapoem. Here's one I wrote for my nephew Jake on his tenth birthday:

"Jacob at Ten"

J umping and jiving like you've got
A nts in your pants.
C an't you sit still? Guess not.
O h well. You're a lively, loving,
B eautiful boy!

Here are some suggestions for writing alphapoems:

- Write the entire spine of your poem down the page at once.
- Let yourself be spontaneous, curious, and open. Overthinking shuts down the process.
- When a word starting with the next letter appears in your mind, take it. You'll make it work.
- If there is internal chatter that you can't write poetry, or this sounds really dumb, just allow it to be there, and put your intention and attention on the form and take the action of writing.
- When I'm feeling confused or stuck, I write alphapoems titled *What's Going On?* Twelve lines later, I nearly always have an answer—and a surprise or two as well.
- Experiment with the A-Z form as well as phrases, words, and names.

We'll practice the series of three again in these Inklings. The topics are all-purpose and useful across many techniques. You could also do a series of three clusters, sprints, or dialogues on this template:

1. *Challenge:* Write about something that is difficult, painful, or stuck.
2. *Best outcome:* Using the same topic, write about the best possible outcome from your perspective. It's fine if it feels out of reach, but don't make it fantastical.
3. *My next step:* Using the same topic, write an alphapoem about action you can take that would move you closer to your best outcome.

Then read all three poems together and reflect. Jack and Caitlyn have donated examples.

It's best to do all three poems at once. Each one will take five minutes or less (if it takes longer, you may be thinking too much), so plan on about twenty minutes for this process.

Inklings

Alphapoem 1: Challenge

Instructions: Bring to mind a current challenge. Describe the challenge in an alphapoem. Write quickly and spontaneously.

Three Feeling Words
_____
_____
_____

C
H
A
L
L
E
N
G
E

Challenge: Jack

Can't believe
How long it's been
Almost a year now since my son
Left home
Love him still, always will
Even if I can't find him to tell him so
Nothing will make me
Get over it
Every day I think of him

Challenge: Caitlyn

Chowing down again
Hating the way I feel
After I eat too much
Like I have no control

Lose my common sense
Energy depleted
Nothing seems to feed it
Groping for the place that is
Enough

Alphapoem 2: Best Outcome

Instructions: What is the best possible outcome? Write another alphapoem that describes it.

B
E
S
T
O
U
T
C
O
M
E

Best Outcome: Caitlyn

Back to my food plan
Every day take medication
Surrounded by healthy snacks
Tomatoes, almonds, celery, plums
Often eat snacks—every three hours
Usual and regular exercise
Treat myself kindly
Come in from the cold
Observe problems with detachment
Mindfully reaching the place that is
Enough

Best Outcome: Jack

Boy, I want you home again
Even if we need to work it out
Son, just come home
Talk to me, we can work this
Out. We can
Unite in our unspoken
Thoughts and feelings.
Communication is the key.
Only you know why you left. Tell
Me—I will listen.
Every day I want you home.

Alphapoem 3: My Next Step

Instructions: Describe a possible next step in your third alphapoem. Remember that you can use Xceptions for Xtra hard letters!

M
Y
N
E
X
T
S
T
E
P

My Next Step: Jack

My son,
You're gone and
Nothing will fill the hole
Except your return. I will
Xamine what I did that broke
Trust

Severed
Ties
Especially the way I was too
Proud to say I was wrong.

My Next Step: Caitlyn

My next step is to
Yield to my better thinking
Not defaulting to habit
Experience shows that
Xploring new paths is not
That hard. It's often
Simply a matter of remembering
To do it. So in
Essence my next step is
Pay attention and take notes.

Reflection

Instructions: Read all three of your poems. What do you see? How do you feel? Are you willing and able to take your next step? When you are complete with your reflection, write three feeling words or phrases that describe how you feel right now. Compare them to the words at the start of the writes. Any thoughts?

Three Feeling Words

_____

_____

_____

Note

1. Although acrostics and alphapoems share an alphabetical structure, "one of the primary differences between acrostics and alphapoems is continuity. In acrostics, the lines tend to be stand-alone, each one responding to the core theme (words or phrases that describe or identify the theme) but independent from one another. In an alphapoem, the lines flow into each other in a way that does feel poetic. In fact, it is often the case that when alphapoems are read aloud, the alphabetical structure falls away and what is left is a piece of writing that can credibly be called a poem." (Adams 2006b)

19

Short Bursts of Radiance

Deborah Ross

The more that you can encourage the brain to default to a responsive mode rather than a reactive mode, the more likely that you will experience the peacefulness, calm, or well-being that recharges your batteries. In this chapter, we're going to head back into the realm of positive experiences—and this time, the task is to relax and allow the positive to come to you and wrap tendrils of well-being around you.

There are only three steps to inviting the brain to rest in the positive:

1. Turn a positive fact into a positive experience.
2. Savor the experience for five, ten, and then twenty seconds. This awareness allows more neurons to fire and then wire together, creating a neural map.
3. Allow the experience to more deeply penetrate, like the warmth of a fire or the sun. This warmth is cued through noticing emotions and/or body/sensory sensations.

All three of these steps can be accomplished in writes of 3-5 minutes! Here's Kay to tell you more.

Kay Adams

We're going to use the series of three again to create three short captured moments of three to five minutes each. In preparation, you'll need a timer that marks seconds (we'll call this the "seconds timer") and minutes (the "minutes timer"), or you may need two timers. You may also want a scent cue. You may take one or two additional minutes of writing any time you desire.

Choose and write down three qualities or values that fill you with well-being. Scan the list below for inspiration, or come up with your own words. We'll call these "word cues."

Qualities and Values

Acceptance	Empowerment	Integrity	Respect
Achievement	Encouraging	Interdependence	Rights
Adaptability	Equality	Intuition	Safety
Adventure	Facilitating	Justice	Security
Affirmation	Faith	Kindness	Simplicity
Appreciation	Family	Knowing	Spirituality
Authenticity	Fidelity	Leadership	Success
Caring	Fitness	Loyalty	Support
Communication	Fostering	Mission	Synergy
Community	Friendship	Motivation	Talent
Compassion	Generativity	Ownership	Tradition
Competence	Generosity	Peace	Truth
Confidence	Happiness	Play	Unity
Connecting	Health	Potential	Validation
Cooperation	Honesty	Power	Values
Creativity	Honor	Prestige	Vitality
Culture	Humor	Productivity	Wealth
Diversity	Influence	Reflection	Wisdom
Dreaming	Inspiration	Renewal	Worth
Empathy	Integration	Research	Yes

SHORT BURSTS OF RADIANCE

For each write, you'll choose a word cue (Step 1) and bring it to mind. You'll recall a powerful positive experience you've had that mirrors the theme, and you'll hold it for up to twenty seconds (Step 2). Then you will write for three minutes by the timer, allowing the experience to deepen by capturing emotions and sensory cues (Step 3).

Complete all three writes, then read and reflect.

CHAPTER 19

Inklings

Short Burst of Radiance #1

Instructions: Set your seconds timer for five seconds. Set your writing timer for three minutes. Bring to mind your first word cue, followed by a positive experience you associate with the word. Take a deep breath (inhale your scent cue if you are using one) and start the seconds timer. Focus awareness on the positive experience for five seconds. Then write for three minutes about the positive experience, focusing on sensory and emotional details. When complete, continue to the second write.

Three Feeling Words

_____

_____

_____

Short Burst of Radiance #2

Instructions: Set your seconds timer for ten seconds. Set your writing timer for three minutes. Bring to mind your second word cue, followed by a positive experience you associate with the word. Take a deep breath (inhale your scent cue if you are using one) and start the seconds timer. Focus awareness on the positive experience for ten seconds. Then write for three minutes about the positive experience, focusing on sensory and emotional details. When complete, continue to the third write.

Short Burst of Radiance #3

Instructions: Set your seconds timer for twenty seconds. Set your writing timer for three minutes. Bring to mind your third word cue, followed by a positive experience you associate with the word. Take a deep breath (inhale your scent cue if you are using one) and start the seconds timer. Focus awareness on the positive experience for twenty seconds. Then write for three minutes about the positive experience, focusing on sensory and emotional details. When complete, continue to the reflection write.

Reflection

Read all three writes and reflect. What do you notice? What surprises you? Where does this land in your body? What was the cumulative effect of short bursts of radiance?

Three Feeling Words

_____
_____
_____

Conversations

Linnea: Despite the different words and cues I chose and the different circumstances I concentrated on for each, I recognized a definite theme that connected the totality of my writing.

I chose Yes, Faith and Pause. I wrote quickly, surprised at what came pouring out. It's finally sinking in that by taking care of myself physically, spiritually, emotionally, I'm so much better in terms of my energy level and nurturing the relationships I have with my family and friends.

Deborah responds: Many people find there is a theme that unifies the three writes. By focusing on the material, holding it, and paying attention by choosing with intention—the message that was floating around can sink in a little deeper. The power of the pen when wielded with attention and intention! Sometimes it seems like reprogramming our emotional GPS rather than just blindly following its direction. In the traditions of the monks who were wired up for study, the loving kindness meditations that are practiced begin with oneself before extending outward.

Michael: In reading my writes, I noticed that the first write reflected a lot of content and facts. It was more detached, objective and initially written from an observer's perspective. This first write was briefer in length and less emotional.

The second write was more personal and reflected more emotion than it did content. I was surprised that feelings superseded facts in this write. In doing this write, I felt like I was back on the scene of what was taking place. Because of this, I felt a reassuring warmth in the mid-section of my body. This write was more precise with feeling and just about the same length as the first write.

The third write, quite unexpectedly, gave rise to the reflection of feelings and meaning. This write felt like it led me to deeper places within myself, which led to questions, which in turn led to a sensation of greater awareness and connection in my body. This greater connection really surprised me. Actually the word *connection* appeared in each one of the writes, but I did not sense this fully until after the last write, which was on simplicity.

Deborah responds: As we drop into the writes, the tendency is for them to deepen. It sounds as though you did a top-down series beginning in your head with thoughts, moving to heart and feelings, and then to a place of integrated meaning. It sounds like all of the layers unified under an embodied sense of connection. Anchoring in your body is a powerful experience that you can draw upon in the future.

Emily: One interesting observation about myself which is actually very power-laden, came through the write about accepting the limits of my influence. What this writing taught me is about the value of acceptance of limits, of my current ego settings, and how lessening the energy around this might be preparing me for a greater emergence in the future. Feels a bit oxymoronic; I need to vanish in order to return. These are important insights for me in my current life. My brain feels changed for having spent the time on these three short writes. I'll be interested to experiment with how to sustain this and groove it deeper.

Deborah responds: I'm noticing how paradoxical the work of transformation often is. Do you remember Chinese finger puzzles? They are woven tubes. You put your forefingers in each end. When you try to pull them out, you get stuck—the harder you pull, the more stuck you get. The secret is to relax and press inward with your forefingers; that frees the tension. Sometimes I think I should just pass them out in workshops and groups as a reminder that we will likely be meeting a lot of paradox along the way. How to groove it deeper? Practice, practice, practice!

20
A Radical Departure

DEBORAH ROSS

Because the personal journal is by its very nature a present-tense, in-the-moment document, the typical journal writer tends not to do much rewriting or editing. Most experienced journal writers seem to share an unspoken philosophy of following the pen and accepting what comes.

Sometimes, though, it is useful to take a radical departure and focus on rewriting and editing. That's what we'll do in this chapter. We'll revisit the concept of negative language that was first introduced in chapter 12, in which we learned that the masking of a negative is not a positive.

This is not only a concept of neuroplasticity, it is also a finding of the research in expressive writing:

> One of the most surprising findings about expressive writing is that the more that people can use positive emotions in their writing, the more they benefit. ... We see positive emotions when words such as love, caring, funny, joy, beautiful, and warmth are used. The degree to which people can use these words ... predicts health improvements after writing. (Pennebaker 2004, 72)

As a quick review of rewriting your masked negatives into positives: if you are working on developing patience and write, "I'm doing really well—I seldom lose my cool or get sharp or snippy," your brain won't take that in as a positive. That statement will register in the deficit category, what you didn't do or say. Think about how you might feel if someone appreciated you for what you didn't do: "I appreciate that you didn't forget to pick up the bread on the way home." Appreciation paired with deficits—ouch!

A positive rewrite of the patience statement might be, "I did really well—I stayed calm, centered, and pleasant."

This isn't editing to improve the writing. This is editing for brain installation. Your new circuitry will thank you!

Kay Adams

Revisit your three "short burst of radiance" writes from chapter 19. See if you notice any masked negatives or qualified positives. When you encounter one, strike it out and rewrite it as a true positive. This may take some time, so be thoughtful about it. Pay attention to whatever bubbles up for you, and come back to reflect when you're complete. If you need to revisit chapter 13, feel free.

Reflection

Instructions: This reflection might take longer than most; allow yourself seven to ten minutes. What was it like to search out the negatives and replace them with positives? Any surprises? Insights? "Aha" moments?

Three Feeling Words

_____
_____
_____

Conversations

Vivien: I noticed that I had several masked negatives. It was powerful to recognize that and then to have the permission to go back and re-write. It felt gleeful, almost like I was getting away with something. The knowledge that it was perfectly fine, even good, to do this was liberating. It was such a beautiful way of thinking differently.

Deborah responds: I'm glad that you had a "gleeful" (what a great word!) experience with this write. It's good to upend things and give ourselves permission to do something different, knowing that it's in service of healing, vitality, resilience, and more.

Shannon: Even when intending to focus on "taking pleasure," my pen was writing things like how good it feels to "not be stressed." I appreciated very much the direction to re-vision the entries through purely the positive. I was able to cull things down to only the enjoyment of the coffee or the morning's birdsong, the effort in my exercising, and the vulnerability at being open to all of life's directions. The revision was welcome in that I could also create an experience to inhabit as a positive statement to go over again and again as a way to strengthen and keep only the part of experience helpful to me.

Deborah responds: Revisioning is a process; everything we notice, whether it's the morning birdsong or our morning coffee, counts. The more success we acknowledge, the easier it becomes to notice. We're developing new habits. It makes sense that the old ones would still show up; that circuitry has been fired more times. Think of embarking down a new path that will become a well-marked trail over time, as the older one recedes with less active use.

Charlotte: This process reminded me of a sermon at church. Our pastor talked about how we tend to talk about peace in negative terms—the absence of fighting or war—instead of in its positive qualities. He compared this to complimenting his wife by saying that she was "the very essence of not-ugly," that her dress "didn't make her look incredibly overweight," and that her cooking "didn't make him feel like being sick." Casting positives in negative terms just doesn't work.

Deborah responds: Your pastor's sermon is a perfect example of how we customarily language the positive as the absence or masking of a negative. His examples really drive home the point that what we think might be positive, most assuredly does not land as positive.

A good way to check this in your own writing is just to ask your body: When I wrote/read that, did I soften or stiffen? Did I expand or contract? Then rewrite until it becomes soft and expansive. Again, this is not an exercise in making it pretty—it is brain installation.

21
Overwriting the Negative

Deborah Ross

From the time that we are young children until adulthood, now defined to be mid-twenties, the brain periodically prunes weaker or under-utilized connections in service of greater operational efficiency, in much the same way that we prune branches in service of the greater health of the tree as a whole. The pruning process is purposeful, and different areas of the brain are pruned at different times; there appears to be a rotational quality to when pruning occurs.

Although active brain pruning ceases in adulthood and is replaced by natural extinguishment as new circuitry is mapped, there is another way to continually modify the brain. This way overwrites the negative by seeking a positive outcome of the experience—a life lesson learned, a positive quality developed—and holding that in the foreground while the negative experience is moved to background.

In other words, you are practicing holding the dual perspective of a stressful or difficult experience from which emerged (or is emerging) a positive outcome, such as a life lesson, a blessing, or the activation of a resource or quality. The stressful experience is subordinate to the positive outcome; the positive has more focus and a higher quality of attention.

Again, findings in experience-dependent neuroplasticity parallel the findings of researchers in expressive writing. In Pennebaker's (2004) four-part writing protocol, the first two writes focus on the difficult experience itself. The third write pulls back and looks at a broader context. In the fourth and final write, attention is placed on synthesizing the first three writes into an action or lesson that can be taken forward: Is there a difficult aspect of this story that can be released? What positive outcome has emerged? What meaningful story about this experience can move into self-application?

In this research model, the difficult story is in the foreground for the first two writes. Then it moves to the background while the foreground is occupied with a focus on positive outcome and moving forward.

Negative material in our awareness has its own neural map, as does the positive. Given the Velcro/Teflon model, we want the positive to be more prominent. Hence, to overwrite the negative we place the positive in the foreground. It helps to saturate it with sensory stimulation, as we have learned to do with scent, novelty, and vivid mental imagery. Meanwhile, the negative is invited to take an observer seat and is acknowledged, but not given the same quality of attention.

Bringing the positive material to the foreground of awareness, while holding the negative in a subordinated background role, creates the conditions for linkage—neurons firing together that will wire together.

When the process of writing and reflecting is over, the negative material leaves awareness and returns to our memory stores. But it will take with it some facets of the positive! When this negative experience is recalled again at a future time, it will likely bring with it those positive facets that wired while the two were held together.

Kay Adams

Once upon a time I got a random email from an old boyfriend who suggested we explore dating again. My curiosity compelled me down a path of predictable folly while my rational self muttered dire warnings. Not unexpectedly it didn't work, for all the same reasons it didn't work the first time we dated. The only difference was that instead of spending a couple of years trying to make it work, I spent a couple of months observing that things weren't working.

For a week or two after we parted ways I was mad at him and mad at myself for not listening to my higher sensibilities. Then one night as I was complaining in my journal I was engulfed by an unexpected wave of gratitude. I had a sudden epiphany that I had the good sense to call it off as soon as I knew it was a repeated pattern; I didn't snooker myself into believing there was something between us that we could build on. I had a marvelous wash of appreciation for the ways I had learned to be honest with myself and others. My pen flew across several more pages, exulting in a powerful sense of freedom and self-determination and gratitude for my own authenticity.

Now when I think of that experience, the focus isn't on the twice-told relationship. That's definitely in the background. The focus is on my relationship with myself and the confidence I feel in my capacity to take care of myself with healthy boundaries, clear communication, and trust in my own knowing. When I recall this experience, I feel powerful and strong. I overwrote the negative.

For your first Inkling, bring to mind an experience that in itself wasn't great but had a positive outcome that has lasted. It's important for this first write that you work with an experience that already feels digested and complete and from which you've harvested a positive outcome. Ideally, there will be no "zing" of regret or unfinished business.

After you've warmed up, we'll move on to a second Inkling in which you take an experience that is still current—not something that is very serious, traumatic or disturbing, but something that is mildly to moderately unpleasant, difficult or challenging—and you'll practice moving it to the background and focusing on bringing a positive outcome to the foreground. When you're done with both Inklings, you'll reflect.

Plan on seven to ten minutes for each of the Inklings, plus another five minutes or so for the reflection. A total investment of twenty to thirty minutes should move you through all three writes.

You may want to bring your anchoring scent into the writes. You'll also need your seconds-timer and your writing timer.

As a last note, if you're having difficulty placing yourself or the situation in the background, try the opening phrase, "Once upon a time." This phrase instantly invokes the distant past.

You might also write the story in the third person; shift the voice from "I" to "he" or "she" and give yourself the role of compassionate narrator. This offers useful detachment and the capacity to see the longer view compared to the first-person perspective.

Inklings

Overwriting the Negative, Part 1

Instructions: Be sure to note three feeling words before you begin. Choose a story from your past that was difficult or stressful at the time, but you found a way to claim a positive experience, life lesson, or quality from it, and that's what you now remember. Inhale your anchoring scent if you choose. Bring the original story to mind, and then move it to the background—present, but not dominant—while you bring the positive aspect of the story to the foreground. Hold the positive story in the foreground for twenty seconds. Then set your writing timer for seven minutes and write the positive story. You may reference the original story for context, but don't dwell on it. Focus on the positive outcome. Extend the writing time if you'd like. When you are complete, move on to the next write.

Three Feeling Words

_____

_____

_____

Overwriting the Negative, Part 2

Instructions: This time, choose a story that is from your recent past or a current story of mild to moderate stress, difficulty, or challenge. Inhale your anchoring scent if you choose. Bring the difficult story to mind, and then move it to the background—present, but not dominant—while you focus on a life lesson, blessing, or quality that this experience has brought forward. What is the silver lining in this cloud? What can you install that will help you shift into a more positive perspective? Hold the positive story in the foreground for twenty seconds. Then set your writing timer for seven minutes and write the lesson or blessing that can come from this experience. Again, you may reference the original story for context, but don't dwell on it. Focus on the positive. Again, extend the writing time if you do not feel complete at seven minutes. When you are finished, move on to the reflection.

Reflection

Read both stories. What do you notice? Where do you feel this in your body? What insights or aha moments did you glean? When you are complete with the reflection, write three feeling words that describe your emotional state at the end of these sequential writes.

Three Feeling Words

_____

_____

_____

Chapter 21

Conversations

Bridget: I chose to examine a difficult work situation, one that I'm already working on overwriting. Several months ago I started challenging my dread and expectations of doom and asking myself, "what would you like it to be, instead?" This seemingly simple switch really turned things around for me.

These writes let me delve deeper into this, and I think solidified the positives I've taken away from a stressful, unfulfilling period of work. For the write, I called to mind the situation, acknowledging the unpleasantness. But then I really focused on all that I've learned, all that this stressful period provided for my growth, how I rose to the challenge, how I faced the situation by "being the change I wanted to see." Taking the time to really focus on this gave me an immense sense of pride in myself, in my abilities to cope, and to actually come out of it a better person. I think when looking back on this period, I will be able to see a brighter side to it now. I prefer it that way. The negative energy surrounding the situation was bad enough; it felt great to attach useful lessons and a sense of accomplishment to something that was so draining.

Deborah responds: I think the piece that we often miss is the attachment of useful lessons and a sense of accomplishment to a situation that is toxic. You've created the linkage so that a whiff of the negative brings up the whole neural map that now includes the positive. This gives you not only something to celebrate, but also something to build on—whether it's deepening the appreciation for your own abilities or giving you practice in how to get more skilled with this process in general. Congratulations!

Sara: I wrote about some new clarity I have reached in a decision I am now navigating in my professional life. A new level of conviction has been achieved in the face of earlier ambivalence but the tone of this was fairly negative. In the course of this write, I have more clearly attached to a positive declaration. The shift is from reaction to response. I believe this is likely to contribute to more confidence and follow-through in my decision as opposed to self-doubt and judgment regarding a reactive stance.

Deborah responds: It sounds like you've made important shifts. Is there something you can do differently, a step you can take? I've been finding that leaning in or leaning toward is an important follow-up to increased clarity; it feeds new circuitry.

Lee: I'm team leader for a project at work, which is new and I'm not that great at it. I'm having a low-grade conflict with a guy on my team who is a buddy of mine. We have beers after work every couple of weeks and argue (friendly) about everything from politics to raising our kids. I don't feel good about being a leader, so that gets in my way of telling him he's not doing what I want for the project.

I didn't have an idea what the positive outcome might be until I parked myself in the background. Then, in the twenty seconds, I realized that he's my friend. I already know how to talk to him. I even confide in him sometimes. All of a sudden I thought, *we can have beers*. I can use what I already know. It gave me a window of possibility that felt real and actionable. I even bullet-listed the points I want to tell him.

Deborah responds: You've described your process of noticing what it is that you already know. You have a good relationship with your buddy, you know where you meet to connect with each other, and you brought in the strengths that come from your role as team leader. Beers and a bulleted list—sounds like you're covering the bases that draw on strengths and provide a matrix for action. Hope it's a successful meeting!

III

OUR LAST COLLECTIVE FIRINGS

22

Reprise

Your Brain Takes the Shape Your Mind Rests Upon

DEBORAH ROSS

In this chapter we'll revisit the My Brain cluster you created in chapter 9 and compare it to see where you are now. There is no particular expectation for this process; stay curious, spontaneous, and open. It's just another snapshot in time.

KAY ADAMS

We're coming to the end of our writing experiment and entering into our last collective firings. Take a moment to consider that you have been a part of a large energy field of people all over the United States and indeed the world who are participating in this process with you. Imagine for a moment the collective power of your intention, attention, and action—your self-directed neuroplasticity. What might that force field be like?

As Deborah said, the writing process for this chapter is the second cluster on My Brain. Let yourself pay attention to what your mind is resting upon these days. Where do you place your attention? Where is the dominant energy? No judgment, just curious, open, observant—write what comes, and then reflect.

CHAPTER 22

Inklings

Reprise: "My Brain" Cluster

Instructions: As you did before, start in the middle of the page with the words My Brain, circled. Then extend thoughts and associations off of other thoughts and associations, allowing the web to spin all over the page. Map out everything your brain is paying attention to these days. What is dominant in your life as you are living it today? When you have completed your cluster, lean back, observe it, then continue with a reflection write to synthesize your findings.

Three Feeling Words

_____

_____

_____

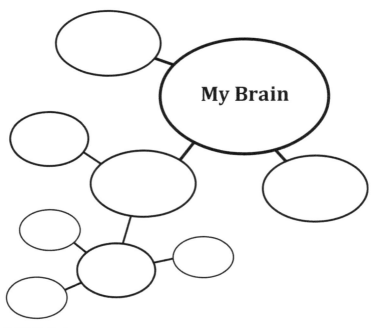

FIGURE 22.1
"My Brain" Cluster

REPRISE

Reflection

Instructions: Review your cluster and notice what you notice. When you are ready to synthesize your learning, write your observations and notes, including any insights or *aha* moments. When you are finished, take one more step. Retrieve your first My Brain cluster and look at the two side by side. What can you discern about the shape your brain is taking, based on this comparison? What differences or similarities do you see? What do you notice?

Three Feeling Words

_____

_____

_____

CHAPTER 22

Conversations

Maggie: My cluster became a series of associations to Velcro and Teflon qualities—common current negativities kicked up by stress and the positive qualities I am choosing to nurture. My first cluster was pretty positive, focusing on the whole brain/mind, but was not as detailed as this one. I realize that tonight I gave myself free rein to write all the negativities as a kind of journaling exorcism, and as a result the Velcro/negative side of the cluster was much more detailed than the Teflon/positive side. Hmmm.

One of the things I found most intriguing and useful in this course (aside from the Velcro/Teflon analogy) was how focusing on positive events/qualities/feelings and saturating the brain/neurons with them increases the "stickiness" of the positive. As I was writing this reflection I remembered the bottle of lavender oil still sitting on my desk from the scent-oriented assignment, and I popped the top, inhaled the aroma, and mentally transported myself to the peaceful creek I'd imagined in that write. I held that experience/imagery/feeling in my mind and body for a minute, focusing on positive qualities. And I feel much more relaxed as a result.

Deborah responds: Do you remember when we spent some time with a part of ourselves that wasn't on board with all this positivity? Giving free reign (the operative word being *free*—no strings attached) to clearing out the negative does create some space. Being intentional about clearing out negatives is different from allowing ourselves to be constantly hijacked. And I'm glad that you remembered your lavender cue.

Bridget: The timing for this content was so helpful to me just now, as I transition from an environment that has been less than fulfilling and not very healthy. Learning to rewire old negativities so that more useful and positive neural firings can happen has made a big difference. I have noticed how much time and work this needs, but that by focusing on these practices for short bursts, regularly, real change can happen.

My brain map has moved from being more cautious and negative in the beginning to more expansive and optimistic. And then, moving on, what has come up for me is the need for stillness. This has come through loud and

REPRISE

clear in the last few weeks in particular. I "hear" this instruction in my brain. *Take on less. Focus on what you need. Be picky.* None of this is negative. It's all for the positive.

Deborah responds: Yes, it is gratifying to realize we can make active positive changes in regular short bursts, relatively quickly! Now that you have a clear message to cultivate stillness, might I suggest that you notice the language you use when you describe it to yourself. If you language what you *will* do in the intentionally positive, you will likely notice more reward from the experience. Remember from chapter 13 that a masked negative (*take on less*) is not a positive. How might you language the clearly positive side of that? If it were me, I'd probably go in the direction of *Be mindful and intentional in choices*. You'll find your own language.

Choose your words carefully. If you go for what you *will* do rather than what you *won't*, and what you *will* do has direct positive associations—then you're already laying down the circuitry toward your desired outcome.

Sara: Two prominent contrasts between my two clusters emerged. The first had more of a "trait" emphasis while the second highlighted "states." The first was more passive in tone while the second was more active. In my reflection, I became aware that the differences were both in the specific content of the two clusters as well as in my relation to the content and language of the two clusters.

Deborah responds: The contrasts between the two clusters—content, language, tone—sound like they represent the learnings and integration of the material. That's pretty great! Keep making the associations—and every time you do, imagine that a new piece of circuitry just got installed!

23
Bridging into the Future

DEBORAH ROSS

What an adventure we've had together! I'm deeply grateful to you, dear readers, for having taken a chance and experienced Your Brain on Ink. It is my hope that you've come away with your neurons crackling with the possibilities of using your journal to cultivate vitality, growth, vision, resilience, joy and all manner of things that call to you in service of a life well-lived.

Well-being is vitally important, and it is contagious (Christakis and Fowler 2011). Now more than ever, in our personal lives and the lives of the communities and world in which we live, it matters that we can skillfully use the instruments at our fingertips to cultivate healing, harmony and hope.

KAY ADAMS

You have brains in your head. You have feet in your shoes.
You can steer yourself any direction you choose.

—*Geisel (Dr. Seuss) 1990*

You've come to the end of the book, and we hope this is only the beginning for you. You're equipped with the tools you need to continue laying down your own neural circuitry for self-directed neuroplasticity: theoretical concepts from the neuropsychology world; concepts about the value of a mindfulness meditation practice, whether that be sitting meditation or a movement meditation such as yoga; and the Journal Ladder and all it contains from the expressive writing world. You've learned and practiced ten journal techniques plus the reflection write. Depending on how long and consistently you've been writing (remember that it takes about six weeks of Brain on Ink writing, minimum one hour a week spread over three to six sessions), you may be well on your way to creating a writing practice.

Next steps? Here's a trail map for the next three to six months.

- Practice, practice, practice. Your journal will document your progress; just keep dating every entry. You'll be able to track your own trajectory toward positive neural change.
- Continue to keep a separate journal for this work for the next three to six months. It often accelerates the process to be able to see the mapping of your writes and especially your reflections. You'll know when it's time to integrate your positive neural writes with your personal process journal; the two will inform each other.
- Add at least weekly (twice weekly is preferred) five-to-ten minute check-ins about daily life to your Brain on Ink journal. Stay observant to any areas in your life where things seem to have more flow, ease, regulation, or integration. Reflect on your successes.
- Your brain loves the wash of emotion that comes with gratitude. Toss in a list of "Three Things I'm Thankful For" whenever you can, remembering to use purely positive language.
- Use this workbook again and again. The writes will hold up over multiple applications. Keep probing your own life stories and harvesting their wisdom.
- If you want to really turbo-charge the process, gather a group of like-minded friends whom you trust, and work through the book together. Become possibility partners for each other. Doing this work in community can be magical, as it is easy to hold for each other what we have a hard time holding for ourselves. That means every person in your group believes in your possibilities just as much as you believe in theirs (when they can't). I've included a handy list of ground rules for group work at the end of this chapter.
- Our wounded, aching world needs conscious citizens and neighbors. As you grow in your own vitality, ease, and resilience, you'll grow in your capacity

to stay calm, kind, and in service under all manner of life circumstances. The ripple effect of this is subtle and strong.

For our last Inklings, we'll use a familiar technique—perspectives—and, in one last chance for novelty, the unsent letter technique (although technically I suppose this is a sent letter—well, you'll see). Of course you'll want your scent cue and your timer handy, and you may want to add some variety with a new pen or by writing at an angle across the page. If you're keyboarding, choose a new font.

Unsent letters are one of the most tried-and-true journal techniques of all. They're great for clarity, catharsis, closure, and communication. Because they're not sent, they can be written without the social filter of whether what you're writing is appropriate. You can be just as raw, honest, sentimental, or churlish as you wish. Your journal absorbs it all.

In the first Inkling, you'll anticipate the future. Then we'll finish with a surprise letter. There are separate reflections for each.

Inklings

Jumping Time: Outcomes

Instructions: Bring to mind a situation in your life that is currently challenging. Imagine it is several months in the future. Pick the date and write it at the top of your page. Imagine that you have continued to practice these methods and are having noticeable positive results. Now imagine these results are reflected in your current challenge and you are seeing results and outcomes. What's different and better? Write for about ten minutes.

Three Feeling Words

_____
_____
_____

Reflection

What do you notice? Where do you feel it in your body? Any surprises? What action steps are suggested?

A Letter From Your Journal

Instructions: Your journal wants to write you a letter. What does it want you to know?

Dear (your name),
What I want you to know is. . . .

Reflection

How did it feel to write and receive a letter from your journal? What do you want to say back, in reflection?

Three Feeling Words

_____
_____
_____

Writing Group Ground Rules: This Group CARES!

Every group benefits from common-sense agreements that will keep the group experience comfortable and accessible for everyone. By agreeing to these ground rules, each individual in the group acknowledges personal responsibility to a shared vision of respect, trust and safety. May we all indicate agreement with these ground rules for our Your Brain on Ink study group?

Confidentiality. We are each empowered to talk about our own experience within this group with anyone. We each agree not to talk about anyone else's experience without the other's specific permission.

Acceptance. Our stories are diverse and rich. They have shaped us. We accept our own stories and the stories of others detached from judgment, critique or the need to fix, resolve, or over-praise.

Respect. We respect the integrity of the group by arriving on time, staying the entire time, and notifying each other if we will miss a session. We respect each other by listening attentively and compassionately. We refrain from interrupting, side-talking, monopolizing, critiquing, or debating. We stay on track.

Empathy. We are all in this together. We care for each other with support and encouragement.

Self-management. We agree to participate at the level that we are able, and we are aware that we have full permission to decline or change any invitation, suggestion or instruction.

References

Adams, Kathleen. 1990. *Journal to the self: Twenty-two paths to personal growth.* New York: Grand Central Publishing.

———. 1994. *Mightier than the sword: The journal as a path for men's self-discovery.* New York: Warner Books.

———. 1998. *The way of the journal: A journal therapy workbook for healing.* 2nd ed. Lutherville MD: Sidran Press.

———. 2006a. Journal therapy: Writing as a therapeutic tool: A training workbook. Brentwood TN: Cross Country Education.

———. 2006b. Alphapoems. In *Writing works: A resource handbook for therapeutic writing workshops and activities,* G. Bolton, V. Field and K. Thompson, eds. London: Jessica Kingsley. 46–50.

———. 2013. Expression and reflection: Toward a new paradigm in expressive writing. In *Expressive writing: Foundations of practice,* K. Adams, ed. Lanham MD: Rowman and Littlefield Education. 1–29.

Adams, Kathleen, Nathan Ohren, and Brenda Hudson. 2015. The 30-day digital journal challenge: A report for helping professionals. Wheat Ridge CO: Center for Journal Therapy.

Amacher, Peter. 1965. *Freud's neurological education and its influence on psychoanalytic theory.* New York: International Universities Press.

Assagioli, Roberto. 2000. *Psychosynthesis: A collection of basic writings.* Amherst, MA: Synthesis Center.

Baldwin, Christina. 2007. *Storycatcher: Making sense of our lives through the power and practice of story.* Novato, CA: New World Library.

Burton, Chad M. and Laura A. King. 2004. The health benefits of writing about intensely positive experiences. *Journal of Research in Personality* 38 (2): 150–163.

Christakis, Nicholas A. and Fowler, James H. 2011. *Connected: The surprising power of our social networks and how they shape our lives.* Boston: Little Brown and Co.

Chung, Cindy K. and James W. Pennebaker. 2008. Variations in the spacing of expressive writing sessions. *British Journal of Health Psychology* 13: 15–21.

Davidson, Richard and Sharon Begley. 2012. *The emotional life of your brain.* New York: Hudson Street Press.

Edelman, Gerald and Vernon B. Mountcastle. 1978. *The Mindful Brain.* Cambridge, MA: MIT Press.

Frattaroli, Joanne. 2006. Experimental disclosure and its moderators: A meta-analysis. *Psychological Bulletin* 132 (6): 823–865.

Geisel, Theodor (writing as Dr. Seuss). 1990. *Oh, the places you'll go!* New York: Random House.

Gottman, John. 1994. *Why marriages succeed or fail.* New York: Simon and Schuster.

———. 1999. *The seven principles for making marriage work.* New York: Three Rivers Press.

Greenberg, Melanie A., Camille B. Wortman, and Arthur A. Stone. 1996. Emotional expression and physical heath: Revising traumatic memories or fostering self-regulation? *Journal of Personality and Social Psychology* 71 (3): 588–602.

Hanson, Rick. 2011. *Just one thing.* Oakland: New Harbinger Publications.

———. 2013. *Hardwiring happiness.* New York: Crown Publishing.

Hebb, Donald. 1949. *The Organization of Behavior.* Hoboken, NJ: John Wiley & Sons.

James, William. 1890 (reprinted 2013). *The principles of psychology: Volume 1.* New York: Cosimo Classics.

Kandel, Eric. 2006. *In search of memory: The emergence of a new science of mind.* New York: W. W. Norton and Co.

REFERENCES

King, Laura A. 2001. The health benefits of writing about life goals. *Personality and Social Psychology Bulletin.* 27 (7): 798–807.

———. 2002. Gain without pain: Expressive writing and self-regulation. In *The writing cure: How expressive writing promotes health and emotional well-being.* S. Lepore and J. Smyth, eds. Washington DC: American Psychological Association.

King, Laura and Kathi N. Miner. 2000. Writing about the perceived benefits of traumatic events: Implications for physical health. *Personality and Social Psychology Bulletin.* 26 (2): 220–230.

Kornfeld, Jack. 2008. *The wise heart.* New York: Bantam Books.

Maguire, E., D. Gadian, I. Johnrusde, C. Good, J. Ashburner, R. Frackowiak, and C. Frith. 2000. Navigation-related structural change in the hippocampi of taxi drivers. *Proceedings of the National Academy of Sciences,* 97: 4398–4403.

Panksepp, Jaak and L. Biven. 2012. *The archeology of mind: Neuroevolutionary origins of human emotions.* New York: Norton.

Penfield, Wilder and Theodore Rasmussen. 1950. *The cerebral cortex of man.* New York: Macmillan and Co.

Pennebaker, James W. 1989. Confession, inhibition and disease. *Advances in Experimental Social Psychology.* 22: 211–244.

———. 2000. Telling stories: The health benefits of narrative. *Literature and Medicine* 19 (1): 3–18.

———. 2004. *Writing to heal: A guided journal for recovering from trauma and emotional upheaval.* Oakland CA: New Harbinger Publications (out of print with New Harbinger; 2013 edition by Center for Journal Therapy, Wheat Ridge, CO).

———. 2013. Foreword, in *Expressive writing: Foundations of practice.* Kathleen Adams, ed. Lanham MD: R&L Education.

Pennebaker, James W., and Sandra K. Beall. 1986. Confronting a traumatic event: Toward an understanding of inhibition and disease. Journal of Abnormal Psychology 95 (3): 274–281.

Pennebaker, James W., Michelle Colder, and Lisa K. Sharp. 1990. Accelerating the coping process. *Journal of Personality and Social Psychology* 58 (3): 528–537.

Pennebaker, James W. and Martha E. Francis (1996). Cognitive, emotional, and language processes in disclosure. *Cognition and Emotion* 10: 601–626.

Pennebaker, James W. and Janel D. Seagal. 1999. Forming a story: The health benefits of narrative. *Journal of Clinical Psychology* 55 (10): 1243–1254.

Poon, Alvin and Sharon Danoff-Burg. 2011. Mindfulness as a moderator in expressive writing. *Journal of Clinical Psychology* 67 (9): 881–895.

Progoff, Ira. 1992. *At a journal workshop*. Los Angeles: J.P. Tarcher.

Rico, Gabrielle Lusser. 2000. *Writing the natural way (15th anniversary edition)*. Los Angeles: J.P. Tarcher.

Schwartz, Jeffrey and Sharon Begley. 2002. *The mind and the brain*. New York: Harper Collins.

Shatz, Carla. 1992. The developing brain. *Scientific American,* 267 (3) 60–67.

Siegel, Daniel J. 2012. *The pocket guide to interpersonal neurobiology*. New York: W.W. Norton.

Smyth, Joshua M. 1998. Written emotional expression: Effect sizes, outcome types, and moderating variables. *Journal of Consulting and Clinical Psychology* 66 (1), 174–184.

Tamura, Hiroshi. 2001 (Winter). Poetry therapy for schizophrenia: A linguistic psychotherapeutic model of renku (linked poetry). *The Arts in Psychotherapy*, 28:5, 319–328.

Whitman, Walt. 1892. Song of myself. In *Leaves of Grass*. The Poetry Foundation, http://www.poetryfoundation.org/poem/174745. Accessed January 5, 2016.

Literature Review

Evidence-Based Research on Expressive Writing[1]

Social science research on expressive writing[2] dates from the mid-1980s. It was then that the first landmark study was published correlating emotional release, or expressive, writing with improvement in overall health functioning (Pennebaker and Beall 1986). Over the next several years, expressive writing studies carved out a method that would hold up across three decades and hundreds of applications. Three or four segments of writing, each taking 15–20 minutes, were typically spaced one day apart.[3] All but a few studies included random assignments of participants to one or more experimental expressive writing groups or to a control group that wrote about deliberately neutral, mundane, or superficial topics. In most of the early studies, experimental writers were instructed to "write deeply" about a topic that was both emotionally difficult and difficult to share. Early on, the data established that writing about both content and affect (as opposed to only about the facts of the story, or only about the emotions experienced) offered the best outcomes, so experimental writers were typically instructed to include both facts and feelings (Pennebaker 1989).

The health benefits from the initial studies were promising indeed: increased immune system function, fewer visits to health care providers for illnesses, lowered blood pressure, improved heart rate, increased antibodies for Epstein-Barr and hepatitis B patients, and other physiological improvements (Smyth 1998). Because of the cathartic nature of writing deeply about difficult, often traumatic, material, the earliest theories conceptualized the expressive writing process as a release valve that depressurized inhibition through the silent "confession" of deeply troubling material. Pennebaker and Beall wrote:

> [Statistical analyses alone do not] convey the powerful and personal nature of the majority of trauma condition essays. One woman wrote about teaching her brother to sail; on his first solo outing, he drowned. The father of a male subject separated from his mother when the subject was about 9 years old. Prior to leaving home, the father told the subject that the divorce was the subject's fault. ... When she was 10 years old, one female subject had been asked to clean her room because her grandmother was to be visiting that night. The girl did not do so. That night, the grandmother tripped on one of the girl's toys, broke her hip, and died of complications during surgery a week later. (1986, 277)

Through the first decade of the writing experiments, it was becoming clear that the arc of the three or four essays seemed to organize thoughts and feelings into a coherent, consistent narrative. Pennebaker and Seagal (1999) wrote:

> [The process of constructing story] allows one to organize and remember events in a coherent fashion while integrating thoughts and feelings. In essence, this gives individuals a sense of predictability and control over their lives. Once an experience has structure and meaning, it would follow that the emotional effects of that experience are more manageable. Constructing stories facilitates a sense of resolution, which results in less rumination and eventually allows disturbing experiences to subside gradually from conscious thought. Painful events that are not structured into a narrative format may contribute to the continued experience of negative thoughts and feelings. (1243)

When understanding is reached, and thoughts are organized into a coherent, cohesive narrative, insight and meaning are accessible. The authors continue:

> The beauty of a narrative is that it allows us to tie all of the changes in our life into a broad comprehensive story. That is, in the same story we can talk both about the cause of the event and its many implications. Much as in any story there can be overarching themes, plots, and subplots—many of them arranged logically and/or hierarchically. Through this process, the many facets of the presumed single event are organized into a more coherent whole. (1250)

Poon and Danoff-Burg (2011) offered mindfulness ("paying complete attention to the experiences occurring presently, in a nonjudgmental way or an accepting stance") as a moderator that might deepen the writer's capacity to create insight and meaning. Study participants completed several inventories, including the Freiberg Mindfulness Inventory, before following instructions to write three times over several days, for twenty minutes each time, about a stressful experience.

> Mindfulness influenced the extent of benefits produced by expressive writing. . . . [A] higher mindfulness score predicted greater change over time in decreased physical symptoms, decreased psychological symptoms, and decreased negative affect, but an increase in sleep quality and positive affect. These findings suggest that people who are more mindful benefit more from disclosing their emotions and thoughts regarding stressful experiences than do those who are less mindful. (890)

Laura King, a professor of psychology at the University of Missouri and a proponent of the positive psychology movement, has conducted several studies that explored factors other than trauma, stress, or illness. In the earliest study, King and Miner (2000) demonstrated that those who wrote about the perceived benefits of a trauma ("the positive aspects and how the experience has benefited you as a person—how has the experience made you better able to meet the challenges of the future?") (233) had the same physical benefits as those who wrote about loss or trauma.

King continued her exploration of writing about self-regulatory topics with a 2001 study comparing the health benefits of writing about trauma with writing about life goals. Her findings:

> Results indicate that writing about life goals is another way to enjoy the health benefits of writing without the emotional costs. Indeed, writing about one's life goals was associated with feeling less upset, more happy, and getting sick less often. . . . [T]he physical benefits . . . were equal to or better than writing about trauma, whereas writing about a traumatic life event also entailed feeling upset and experiencing lowered mood. . . . It may be possible to enjoy the benefits of writing without necessarily writing about trauma at all. (805)

Burton and King (2004) asked experimental subjects to write about nonemotional topics or intensely positive experiences (IPEs) for 20 minutes each over three consecutive days. IPE writers showed increased positive mood, LIWC analyses revealed that their language reflected positive and insightful aspects, and everyone in the IPE portion of the study remained healthy. However,

> *none* of the mediational analyses were successful in identifying the underlying mechanisms of this effect. Thus, this study, like others using the disclosive writing paradigm, indicates that writing "works"—i.e., it is associated with superior health—but does not provide a strong indication for why this is so (King 2002).[4] . . . This study broadens our appreciation for the "healing power of writing." Writing about life experience may have more general implications than have previously been explored. Furthermore, our results suggest that the discussion about the potential mechanisms of the writing effects should be broadened to include a variety of processes that are not exclusive to confronting life events. (159–161)

Notes

1. Previously published in Adams 2013.

2. In the research model, "expressive writing" generally refers to the disclosure or release of emotions, traumatic memories, secrets, and other disempowering thoughts and feelings.

3. The time interval between writes does not appear to be a moderator. A meta-analysis by Frattaroli (2006) did not find a difference in effects as a function of the spacing of writing sessions when intervals between writes were longer than 24 hours. A study by Chung and Pennebaker (2008) determined that shortened intervals (completing one write per hour for three hours, or even three writes in one hour) "is as effective as the traditional once-per-day approach" (16).

4. "Two strong conclusions can be made with regard to the benefits of writing. First, expressive writing has health benefits. Second, no one really knows why" (King 2002, 119).

Acknowledgments

DEBORAH ROSS

The biggest thank you is to Kay Adams, my mentor, co-author, co-teacher and source of inspiration and knowledge in all things related to expressive writing. When I first proposed the development of a curriculum called "Your Brain on Ink" for her curriculum design class, she responded with a resounding yes. Her guidance and wisdom have been invaluable both as a source of grounding in the therapeutic writing field and encouragement to explore the soaring new frontiers in neuroscience. The rest is an exciting history of integration and collaboration that continues to be a challenging, circuits-firing, stimulating work in progress.

Thank you to Salma Abugideiri and Trisha Stotler who have been on this journey with me. Salma suggested that we do Kay's training on journaling and as my colleague, Salma has championed the unfolding and birthing of this new stage of my professional life. Trisha, as my meditation teacher (I took her Mindfulness Based Stress Reduction class twice) helped me revive my meditation practice which has led to wonderful conversations and collaboration at this intersection of mindfulness practices, the brain and the pen.

I also wish to thank Dr. Dan Siegel, the staff and my fellow students at the Mindsight Institute. The three-year program in Interpersonal Neurobiology deepened and changed the way I worked as a therapist and led me to consider how writing might facilitate integration across multiple domains. Now I can't look at my hand without smiling and seeing a model of my brain.

Thank you to all of my teachers, colleagues, students, and the staff at the Therapeutic Writing Institute. Your feedback, questions, insightful comments, support, encouragement and generosity have been a platform and a model for how to be a nurturing and expansive writing community. It is an ongoing pleasure and gift to be a part of this Institute, first as student and now as alumni and faculty.

My clients and journaling students have also lent their voice via the pen to my growth as a journal therapist. Their questions, engagement with the process, lightbulb moments and laughter have kept the work vibrant and exciting.

Finally, to my family, friends and colleagues who have been living "the book" alongside me, a special thank you for your support and all the things that you picked up as I was holed up. Chuck, Sam, Sanj, Liz, Ben, Ibrahim, Aminah and Pam—thank you!

Kay Adams

Thanks to my collaborator and coauthor, Deborah Ross, who has taken therapeutic writing to a whole new paradigm with this work. I love teaching, working, and writing with you, and I can't wait to discover what we'll fire up next!

Deep gratitude to the series' editorial review board, Dr. Peggy Osna Heller, Dr. Vanessa Furse Jackson, Linda Barnes, and the memory of the late Dr. Robb Jackson, who have devoted endless hours of service over the span of ten books to keep me aligned with best practices.

And much gratitude to our colleagues who read this manuscript (sometimes several times!), and whose mindful reading and thoughtful notes helped me see more deeply: Susan de Wardt, Leia Francisco, Wendi Kaplan, Nancy Scherlong, Kate Thompson, and Mary Reynolds Thompson, and to Marta Hedde for last-minute graphic help. We get by with a lot of help from our friends.

Many thanks to all the clients and students whose sharing makes this material come alive.

This is the last of ten books in the It's Easy to W.R.I.T.E. expressive writing series. I am very grateful to have been entrusted with a major body of work, and I have had a dream editorial team at Rowman & Littlefield: my editor, Sarah Jubar; associate editor, Carlie Wall; book designer, Maria Kauffman; and all the production team. Thanks to marketing director Dean Roxanis for working with me to bring these books to their audience. I also offer bountiful thanks to Nancy Evans, the original acquisitions editor who suggested and then signed me to the series.

My family continues to be my anchor, and once again they have cheerfully fed me, hauled me away from the computer for movies and walks, listened patiently to my brain dumps, and supported me completely. I love you each and all.

Thanks be to God, from whom all blessings and books flow.

About the Authors

Deborah Ross, LPC, CJT, is a psychotherapist, certified journal therapist, teacher, and writer. She is the creator of the original curriculum, Your Brain on Ink. Deborah is fascinated by the intersection of neuroscience, mindfulness-based practices, and writing. She studied with Dr. Dan Siegel, successfully completing his three-year program in Interpersonal Neurobiology through the Mindsight Institute. She also successfully completed the three-year Certified Journal Therapist program at the online Therapeutic Writing Institute founded by her coauthor, Kathleen Adams.

Based in the Washington, D.C., area, Deborah teaches journal writing in healthcare programs and in a year-long Bringing Mindfulness to Life meditation program, and she is currently creating a program for fiber artists wishing to use their journals in service of creativity. Deborah continues to focus on the development of writing strategies for positive brain change through her faculty role at the Therapeutic Writing Institute, where she teaches her Your Brain on Ink curriculum as well as core theory and psychological awareness classes. She nourishes all of this cerebral activity by extended time outdoors.

Kathleen Adams, LPC, PTR, is a licensed professional counselor and registered poetry/journal therapist who has pioneered the field of expressive writing in counseling, therapy, education, and community service since 1985. She is the founding director of the Center for Journal Therapy, Inc. in Denver,

Colorado, and its professional training division, the online Therapeutic Writing Institute. In 2013 she created the Journalverse, an online membership community for journal writers and facilitators worldwide.

She served on the board of directors of the National Association for Poetry Therapy for twelve years, including a term as president from 2001–2003. Kathleen has also served on the board of the International Federation for Biblio/Poetry Therapy, the credentials agency for the field, where she helped write policy for best-practice training standards.

Kathleen is the author/editor of twelve books on writing as a tool for healing, growth, and change, including six titles from this series. From 2012–2016 she served as series editor for the It's Easy to W.R.I.T.E. expressive writing series and brought ten books into form, of which this is the last.

Milton Keynes UK
Ingram Content Group UK Ltd.
UKHW041259100923
428418UK00001B/91